Hypothalamic Control of Pituitary Functions. The Growth Hormone Releasing Factor

The Sherrington Lectureship

In 1948, on the occasion of the ninetieth birthday of Sir Charles Scott Sherrington, O.M., G.B.E., M.D., F.R.S., the Council of the University of Liverpool resolved to institute a Lectureship in recognition of his distinguished contributions to Physiology and Medicine and of his association with the University as George Holt Professor of Physiology from 1895 to 1913. The appointment to the Sherrington Lectureship is made biennially by the Council on the joint recommendation of the Faculties of Medicine and Science.

The Sherrington Lectures

The following Lectures have been delivered and, except as indicated, are published by Liverpool University Press.

I *Sensory Integration*, by E. D. Adrian, O.M., M.D., F.R.C.P., F.R.S., Professor of Physiology in the University of Cambridge.

II *The Frontal Lobes and Human Behaviour*, by John F. Fulton, O.B.E., M.D., D.SC., LL.D., Sterling Professor of the History of Medicine, Yale University.

III *The Invasive Adenomas of the Anterior Pituitary*, by Sir Geoffrey Jefferson, M.S.(LOND.), F.R.C.S., F.R.S., Emeritus Professor of Neurosurgery, University of Manchester.

IV *Sherrington: Physiologist, Philosopher and Poet*, by the Right Honourable Lord Cohen of Birkenhead, M.D., D.SC., LL.D., F.R.C.P., F.A.C.P., F.F.R., F.S.A., J.P., Professor of Medicine in the University of Liverpool.

V *The Excitable Cortex in Conscious Man*, by Wilder Penfield, O.M., C.M.G., LITT.B., M.D., D.SC., F.R.C.S., HON.F.R.C.P., F.R.S., Director, Montreal Neurological Institute, McGill University, Montreal, Canada.

VI *Visual Pigments in Man*, by W. A. H. Rushton, SC.D., F.R.S., Reader in Physiology in the University of Cambridge.

VII *The Conduction of the Nervous Impulse*, by A. L. Hodgkin, SC.D., F.R.S., Foulerton Research Professor of the Royal Society, University of Cambridge.

The Sherrington Lectures XVIII

Hypothalamic Control of Pituitary Functions. The Growth Hormone Releasing Factor

---

Roger Guillemin

Laboratories for Neuroendocrinology,
The Salk Institute for Biological Studies,
La Jolla, California, USA

LIVERPOOL UNIVERSITY PRESS

Published by
LIVERPOOL UNIVERSITY PRESS
PO Box 147, Liverpool L69 3BX

ISBN 0 85323 185 0

First published 1986

British Library Cataloguing-in-Publication Data
Guillemin, Roger
Hypothalamic control of pituitary functions: the growth hormone releasing factor.—(The Sherrington lectures; 18)
1. Hypothalamic hormones
I. Title II. Series
612′.492 QP572.P5

ISBN 0–85323–185–0

Text set in 10/12 pt Linotron Times
Printed and bound in Great Britain
at The Bath Press, Avon

Contents

Acknowledgements

Research supported by program grants from the National Institutes of Health (HD-09690 and AM-18811) and a grant from the Robert J. Kleberg, Jr. and Helen C. Kleberg Foundation. The senior investigators involved in the research dealing with the isolation, characterization and synthesis of GRF, studies on its mechanism of action, are all co-authors of the pertinent publications as quoted throughout the text. At the Salk Institute, staff members involved were P. Böhlen, P. Brazeau, F. Esch, N. Ling, W. Wehrenberg, F. Zeytin, B. Bloch, C. Mougin; in collaboration from the Roche Institute for Molecular Biology: R. Bhatt, K. Collier, P. Gage, B. Hoffman, P. Lomedico, J. Monahan, M. Poonian, U. Gubler; at the Sloan-Kettering Cancer Center: F. C. Bancroft. I am also happy to acknowledge the collaboration of the technical staff of the Laboratories for Neuroendocrinology: B. Alford, D. Angeles, F. Castillo, K. Cooksey, T. Durkin, D. Fuller, R. Klepper, D. Lappi, D. Martineau, M. Mercado, B. Phillips, M. Regno, R. Schroeder, K. von Dessonneck. The services of the secretarial staff of the laboratories in typing and handling the manuscript are also highly appreciated.

Tables

Figures

Introduction

The lecture delivered on 11 April 1984 dealt specifically with the technical problems of the isolation and structural characterization of the hypothalamic growth hormone releasing factor, the latest of these hypophysiotropic factors to be identified (about one year prior to the lecture), and with its implications in physiology and clinical medicine. Because it has not been presented to any extent in any of the volumes of this series, I thought that it would be appropriate to precede the text corresponding more precisely to the lecture as delivered, with an opening chapter that would, though in somewhat general terms, present the overall concept of the neurohumoral control of adenohypophysial function, thus placing the lecture on the new knowledge related to growth hormone releasing factor in its proper physiological context.

Neuroendocrinology as a whole, as we see it now, is remarkably in accord with Sherrington's concept and ideas of the *Integrative action of the nervous system*. Indeed, in the opening chapter of his 1906 volume, Sherrington considers already 'chemical aspects' mediated by the nervous system as part of his concept of its integrative functions. Besides a couple of examples dealing with functions of the gastrointestinal tract, Sherrington writes: 'Integration also results from *chemical* agency. Thus, reproductive organs, remote from one another, are given solidarity as a system by communication that is of chemical quality; lactation supervenes *post partum* in all mammary glands of a bitch subsequent to thoracic transsection of the spinal cord severing all nervous communication between the pectoral and the inguinal mammae (Goltz).' That is classic endocrinology. Had Sherrington added that milk letdown in all mammae does take place upon sucking of the thoracic nipples, he would have described the typical neuroendocrine reflex, i.e. release of oxytocin, followed by its circulation-mediated peripheral effects on the mammary smooth muscle contraction.

The opening chapter of this text on neuroendocrine relationships will present some unpublished diagrams and photographs of the hypothalamo-hypophysial portal system, a crucial anatomical structure in these functional relationships, due to the courtesy of Professor Henri Duvernoy, a distinguished colleague and long-time friend, from the Department of Anatomy of the Medical School, University of

Besançon, France. The text of the lecture, dealing primarily with the isolation, characterization and physiology of the hypothalamic growth hormone releasing factor, borrows heavily from the text written in collaboration with P. Böhlen, P. Brazeau, F. Esch, N. Ling, W. Wehrenberg, F. Zeytin for *Recent Progress in Hormone Research*, **40,** 233–86, 1984 because after all, there is, as it should be, only one way to express or describe one thing or another. The text from *Recent Progress in Hormone Research* has, however, been totally updated when novel information has been available from the time it was first written. Thus, the text sent to the publisher, as here, is an up to date review of the status of the question *in re* growth hormone releasing factor. It covers an astounding amount of new data generated in less than two years after the primary structure of this elusive molecule was established, thus completing the search for all hypothalamic hypophysiotropic peptides corresponding to known pituitary hormones that was started in the early 1950s.

Neuroendocrine Interrelations—The Case of the Growth Hormone Releasing Factor

A generation ago we were taught that the anterior pituitary gland commands all other peripheral endocrine glands (possibly with the exception of the parathyroids and perhaps of the pancreas) and that a subtle reciprocal equilibrium between concentrations of circulating 'peripheral' hormones and quantities of the corresponding pituitary hormones maintained endocrine homeostasis (feedback or push-pull theory). Thus were explained the experimental compensatory hypertrophy of peripheral endocrines following unilateral ablation, the development of goiter, the atrophy of the contralateral adrenal in the presence of a secreting unilateral adrenal tumor, and so forth. Then followed a twilight period in which more and more aspects of this construction lost their clarity and heuristic value as it was realized that there were circumstances in which the simple pituitary-target organ functional relationship was not exclusively operative. For instance, some novel explanation was needed for the excretion in the urine of large quantities of corticoids during prolonged stress such as surgical aggression or even strenuous exercise in human subjects, an observation which was incompatible with the eucorticoidism theory postulated by the feedback system between pituitary and adrenal cortex secretion. Similarly, the extreme rapidity of the changes in pituitary ACTH secretion was realized upon exposure to acute or sudden stress, with increases in plasma concentration of pituitary hormones taking place in a few minutes with no evidence of preceding fall in plasma concentration of the peripheral hormones, which could have triggered the feedback system. Moreover, it was becoming abundantly clear that lesions of the ventral hypothalamus, either experimental (by stereotaxic placing of electrodes) or as the result of some pathology (cysts, tumor, traumas) would produce various syndromes of pituitary dysfunction such as inhibition of stress-induced release of ACTH, permanent diestrus, permanent estrus with an ovarian picture reminiscent of the Stein-Leventhal syndrome, testicular atrophy, etc.

Today we know that the proximal center of control of the adenohypophysial functions is to be found in the hypothalamus, that the feedback relationships between peripheral hormone levels and adenohypophysial secretions can be over-ridden by hypothalamic

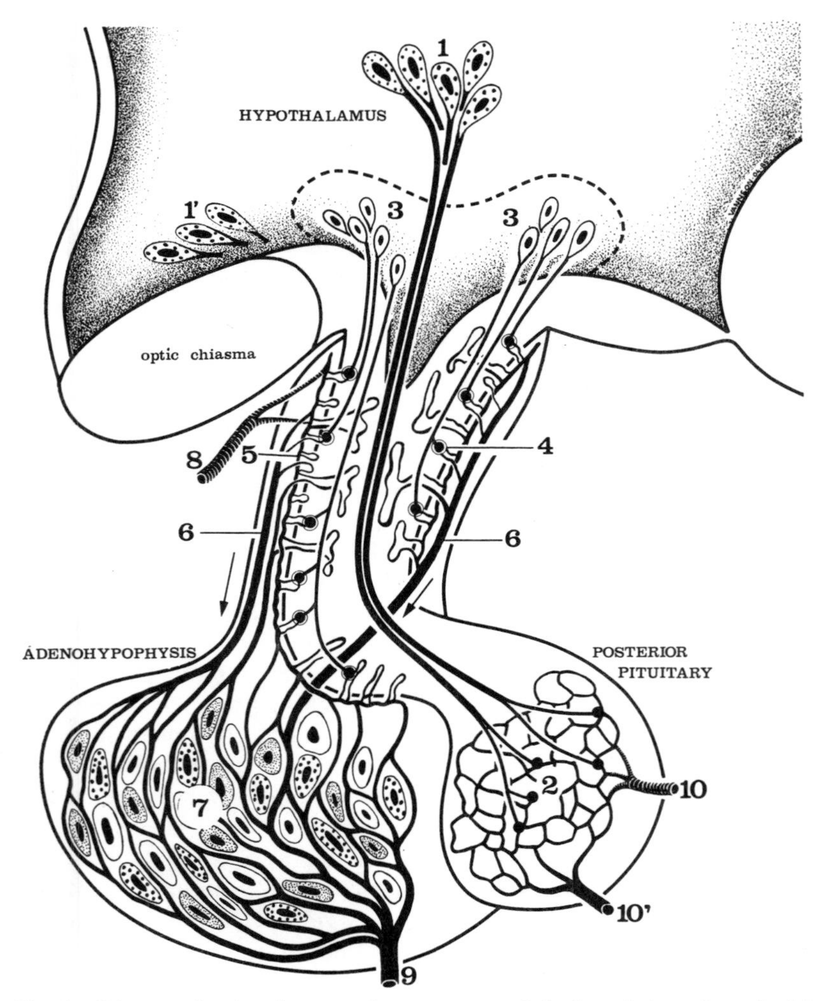

Fig. 1a. Diagram showing the general arrangement of the hypothalamo-hypophysial connections in man (median sagittal section)

—*Hypothalamo-posthypophysial connections:*
1 and 1′ origin of the hypothalamo-posthypophysial tracts (paraventricular and supraoptic nuclei).
2. Endings of the hypothalamo-posthypophysial tracts within the posterior lobe.

—*Hypothalamo-adenohypophysial connections:*
3. One of the hypophysiotropic nuclei of the hypothalamus (nucleus arcuatus).
4. Axon terminals within the median eminence in contact with the capillaries of the primary plexus of the hypophysial portal system.
5. Primary plexus.
6. Portal vessels (arrows indicate the direction of blood flow)

influence, and that the hypothalamic control over the pituitary functions is exerted through the secretion of hypothalamic hormones or releasing factors, which may be called hypophysiotropic substances or hormones.

The purpose of this short review will be to describe briefly the principal tenets of the current concepts in neuroendocrinology without going into any of the techniques that were involved in acquiring the pertinent knowledge. Key references to several reviews or specific articles will be given for the reader who might want to go into the mechanisms of how these were obtained.

There is general agreement (Green, 1951; Harris, 1955) that connections between the hypothalamus and the anterior pituitary are not provided through tracts of nerve fibers (as in the case of hypothalamus to posterior pituitary), but rather through a vascular link in the form of a system of portal vessels. The primary plexus of this system of vessels is to be found in a contact area between ventral hypothalamus and pituitary stalk called the median eminence (Fig. 1). Axon terminals from the neurosecretory cells of hypothalamic nuclei come into close contact with the capillaries of this primary plexus, and there is passage of hypothalamic neuro-humoral factors, the hypothalamic releasing factors, into the blood stream of these capillaries. Hypothalamic releasing factors for one pituitary hormone or another can be measured quantitatively by radioimmunoassays in aliquots of portal blood (pituitary stalk blood). The substances are distributed through collecting veins in the pituitary stalk into the adenohypophysial parenchyma by the secondary plexus of capillaries of this hypothalamo-hypophysial portal system. The names of Geoffrey Harris and John Green will remain associated with their early proposal of such a neuro-humoral control of adenohypophysial functions. Precise localizations of the origin of a given releasing factor in terms of the neurons that synthesize and release it are now well recognized through the use of the methods of immunocytochemistry. It is quite remarkable that these localizations of the hypophysiotropic neurones correspond rather well with anatomical structures described earlier by neuroanatomists (paraventricular n., supraoptic n., n. ventromedialis, n. arcuatus, etc.). Indeed, specific neurons for one releasing factor or another

7. Secondary plexus in contact with the adenohypophysial cells.
8. Superior hypophysial arteries.
9. Drainage of the adenohypophysis.
10. and 10′ inferior hypophysial arteries and veins.

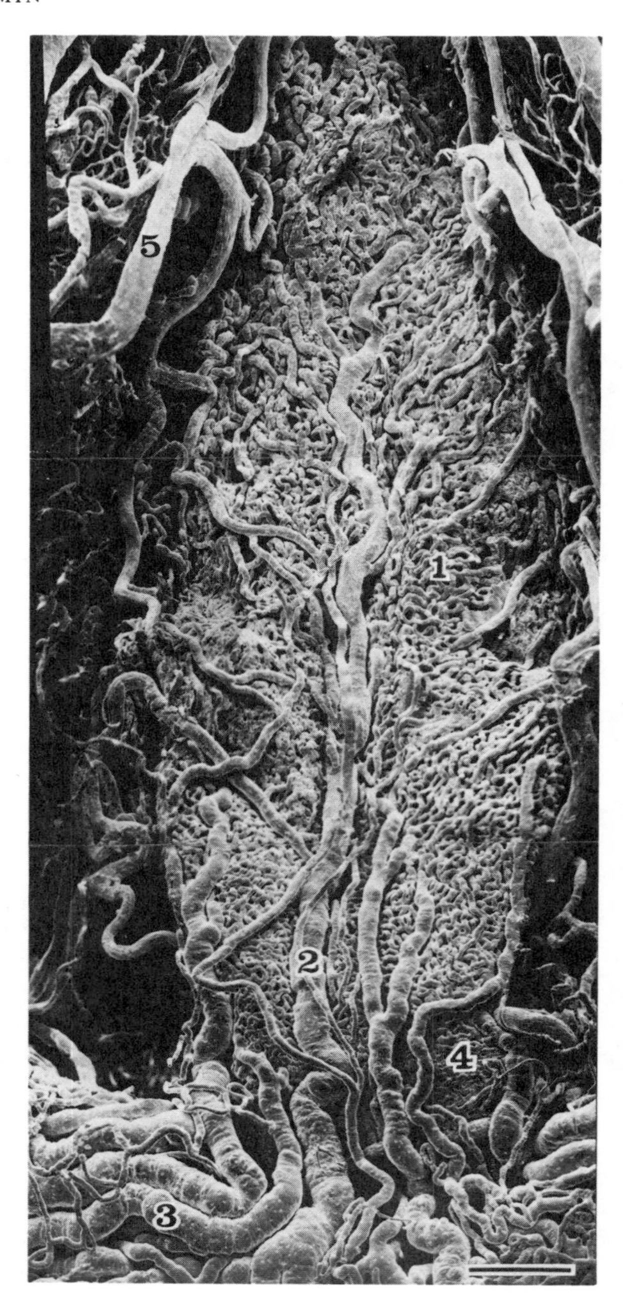

Fig. 1b. Scanning electron microscope; bar = 200μ
Ventral view of the hypophysial portal system (rat).
1. primary plexus covering the median eminence.
2. Portal vessels.
3. Secondary plexus within the adenohypophysis (proximal part).
4. Pituitary stalk.
5. Superior hypophysial arteries.

can actually be shown to be located within subpopulations of the neurones of these (anatomically defined) nuclei, usually within well-defined boundaries.

When the search began in 1955 for hypothalamic releasing factors, the hypothesis which directed the early attempts at their chemical extraction and purification proposed that the hypophysiotropic substances of hypothalamic origin would most probably be of polypeptidic nature, as were the other two recognized products of neurosecretion, oxytocin and vasopressin, isolated, characterized and synthesized (oxytocin) for the first time by Du Vigneaud and his collaborators in 1952. The isolation and characterization of the hypothalamic releasing factors proved to be unusually difficult in view of the minute quantities in which they are found in each fragment of hypothalamus (of ovine, bovine, porcine and murine (rat) origin) used as starting material. Indeed, it is now realized that a fragment of hypothalamic tissue contains no more than 1×10^{-12} to 10^{-10} mole of each of these peptides. Thus several millions of these hypothalamic fragments had to be collected, dissected and extracted in the laboratory to yield quantities of releasing factors meaningful for chemical characterization. Recent developments in the technology of separation of peptides (HPLC) and of microsequencing (gas phase sequencer) have improved the yields of such endeavors by several orders of magnitude. The techniques of molecular biology, with the cloning of the message for any suspected biologically active peptide along with monoclonal antibodies inhibiting the biological activity, allow one even to get the complete primary structure of any one of such molecules nowadays with no more than a few fragments of the pertinent tissues.

In their proposal of a humoral hypothalamic control of adenohypophysial secretions, Green and Harris had left open two possibilities: 1) that the hypothalamic humoral mediator would be a single substance—the pituitary secretion being thus either of all hormones simultaneously, or, if eventually shown to be solitary for one hormone or another, that the specificity would be due to a modulation of peripheral origin (steroids, thyroid hormones, etc.); or 2) that there would be several hypothalamic neurohumors, each one being particularly involved in the secretion of one pituitary hormone. Subsequent physiological studies yielded results best explained by a multiplicity of hypothalamic hypophysiotropic factors.

This latter proposal became an inescapable conclusion when TRF (thyrotropin releasing factor) was characterized in 1969 (Burgus *et al.*, 1969; Nair *et al.*, 1970): TRF would stimulate exclusively the secretion of thyrotropin (TSH), not of ACTH (adrenocorticotropin), or GH (growth hormone) or the gonadotropins (LH and FSH). This remains true even though it was shown later that TRF can stimulate the secretion of prolactin (PRL) along with that of TSH (Tashjian *et al.*, 1971). Indeed the ensuing years saw the characterization of LRF (Matsuo *et al.*, 1971; Burgus *et al.*, 1971), that stimulates the secretion of both gonadotropins LH and FSH, and of CRF (Vale *et al.*, 1981), that specifically stimulates the secretion of ACTH and β-endorphin.

It only remained to characterize a growth hormone releasing factor, GRF. There were many efforts to this end for almost 20 years. Based on questionable bioassays, they yielded results that ultimately had to be recognized as artefacts (a fragment of the β-chain of hemoglobin—Schally *et al.*, 1971; Veber *et al.*, 1971; an inactive tetrapeptide—Yudaev *et al.*, 1973; never fully characterized peptides—Wilber *et al.*, 1971; Nair *et al.*, 1978, etc.). These efforts also led to the recognition and eventual characterization, in extracts of the hypothalamus, of somatostatin, a powerful inhibitor of the secretion of GH (Krulich *et al.*, 1968; Brazeau *et al.*, 1973).

With knowledge of the pitfalls of the bioassays and efficient ways of removing the interfering somatostatin from the crude hypothalamic extracts (affinity chromatography, gel filtration) we started anew to look for GRF in 1978. Rapidly we showed that it was possible by gel filtration on Sephadex G75 or G50 to separate all forms of bioactive somatostatin (SS-14, SS-28) from a single zone of the effluent containing GRF activity, based on an *in vitro* assay system measuring by RIA the secretion of ir-GH by rat pituitary cells grown as a monolayer. Almost as rapidly it became obvious that we had, in each hypothalamic fragment, only minute amounts of GRF, probably no more than 10–50 fmol, assuming a potency for GRF to be of the order of that of the already characterized hypothalamic releasing factors.

I then decided that we should investigate in parallel another possible source of GRF, namely those rare tumors, mostly islet cell adenomas or carcinomas known in patients to accompany a full-blown syndrome of acromegaly in the absence of a pituitary adenoma (see Leveston *et al.*, 1981, for a review of such cases). Indeed, from early studies by Frohman *et al.* (1980) the GRF activity recognized by these authors in the extract of one such tumor appeared to have chromatographic behavior very similar, if not identical, to that of the material we were purifying from hypothalamic extracts.

At first, clinician colleagues holding such GRF-secreting tumors were not anxious to part with aliquots of the GRF-secreting tissues for us to attempt to isolate GRF from them. Then, in 1981, a 200-mg fragment of a lung tumor which had caused acromegaly was obtained from B. Scheithauer at the Mayo Clinic. The amount of tissue obtained was so small that we decided to keep it frozen for use at some later date when we would have more refined technology for isolation and sequencing than what was available at that time; attempts to grow GRF-secreting cells out of that tissue were unsuccessful (two years later we received 57 grams of the same tumor which unfortunately had been kept in formalin; we could not find GRF activity in it). In August 1981 Fusun Zeytin joined us from the laboratories of Armen Tashjian at Harvard, bringing with her about 400 mg of an islet cell tumor which had caused acromegaly in a patient (Ms B. G.) of Michael Thorner, University of Virginia. The tissue had been sent by Michael Thorner to Armen Tashjian in the hope of growing GRF-secreting cells out of it, in parallel with similar efforts conducted by Michael Cronin at the University of Virginia (no group succeeded in that task). With the generous agreement of both Armen Tashjian and Michael Thorner, we extracted what remained of that tumor-aliquot and isolated ca. 100 picomoles of a single GRF entity for which we obtained an amino acid composition on 24 October 1981. No sequence data were attempted, since our then available sequencing equipment would not have had the necessary sensitivity. At that time, Michael Thorner and W. Vale agreed to provide us with an additional 5 g of that tumor, now in the hands of W. Vale. Again that tissue was kept in liquid nitrogen, untouched, until we would have high-resolution sequencing capability, which was now expected for February–March 1982 through the delivery of one of the prototypes of the gas-phase sequencer instrument developed by Hood, Hunkapiller and collaborators (Hewick *et al.*, 1981) and to be manufactured by Applied Biosystems, Foster City, California.

In September 1981, I gave a lecture at the old Faculté de Médecine in Paris, in one of the plenary sessions of the annual meeting of the French Société d'Endocrinologie. The title of the lecture was 'Evidence for a central nervous system control of the secretion of growth hormone' and discussed primarily physiological and pharmacological data and also some clinical studies particularly relating sleep and GH-secretion. In that lecture I mentioned the existence of the tumors as ectopic sources of GRF, a possibility, though rare (and still to be proven), not be overlooked in the diagnosis of the etiology and surgical treatment of acromegaly without any obvious pituitary adenoma, quoting the recent case of Thorner. I also explained my interest in obtaining tissues of any such tumor as a possible source of GRF. A couple of months later I received a letter from Dr Geneviève Sassolas of the Alexis Carrel University in Lyons, France, describing a patient (Mr T. W.) whose history and case report fitted the clinical picture of an ectopic GRF-secreting tumor. I wrote back with a couple of questions and suggestions and on 31 March 1982, a second letter from Geneviève Sassolas describing new clinical findings confirmed the presence of a tumor of the pancreas in that patient. I immediately contacted her in Lyons by telephone. The patient was scheduled for surgery on 16 April. On 13 April, Fusun Zeytin left for Lyons equipped to collect tumor tissue so as to grow GRF-secreting cells out of it (another unsuccessful attempt, as it turned out) and to organize things with the surgical team of Dr Christian Partensky to obtain the bulk of the pancreas tumor in optimal conditions, the tumor to be diced in liquid nitrogen within minutes of surgical removal. Fusun returned to the laboratories on 18 April with the bulk (ca. 200 grams) of two separate tumors both found in the pancreas of the patient. By 21 April, small aliquots of both tumors had been extracted and filtered on G-75 by Peter Böhlen and we knew from the bioassays by Paul Brazeau that both tumors contained large quantities of bioactive somatostatins, with minute amounts of GRF activity in one, but large amounts in the other (5-10,000 times more GRF activity than that found in an equivalent weight of hypothalamic extracts). Two aliquots of 7 g each of that GRF-rich tumor were processed; three peptides with GRF activity were obtained in homogeneous forms on 10 May and 17 May; on 22 May Fred Esch completed the first amino acid sequence of the one peptide with GRF activity that was present in largest amounts. On 28 May, Nicholas Ling had reproduced that sequence by solid-phase synthesis and purified it to homogeneity on 1 June. On 4 June, Paul Brazeau showed that the synthetic replicate had, *in vitro*, the full biological activity of the native material. We now knew that we had isolated and characterized GRF or a biologically active fragment of it (since we knew that we also had a larger molecule, also bioactive) and had reproduced it by synthesis; the sequence of events had taken less than a month and a half. *In vivo* activity was shown on 6 June by Bill Wehrenberg by *i.v.* injection of the synthetic replicate into rats anaesthetized with pentobarbital. These results were announced at a poster session at the meeting of the Endocrine Society on 16 June, though the primary structure was not shown. The primary structures of the tumor-derived GRF (hpGRF-44) and of two biologically active fragments (hpGRF-37, hpGRF-40) were reported in *Science* **218,** 585, 5 November 1982; by the end of the year we had published a series of papers describing the results of *in vitro* studies on the mechanism of action of GRF, *in vivo* studies including the use of a monoclonal antibody against rat hypothalamic GRF, localization of GRF neurons in the brain of primates, synthesis of fragments and analogs, etc.

From the amino acid composition obtained in October 1981, we knew that the GRF peptide isolated from the first 400 mg sample of the tumor obtained from Michael Thorner was closely related to one of the molecules isolated from the Lyons tumor. Processing of the 5 g fragment of that tumor revealed a single molecule with GRF activity. Fred Esch determined the primary structure of that material to be identical to that of the hpGRF-40 characterized earlier from the other tumor. This was reported in *Biochemical and Biophysical Research Communications* **109**, 152, 16 November 1982. In the 18 November 1982 issue of *Nature* (**300**, 276) Rivier *et al.* reported the structure of the GRF peptide they had isolated and characterized independently of our own efforts from another fragment of the Thorner tumor, also as that of hpGRF-40.

The same group (Spiess *et al.*, 1983) reported a structure for hypothalamic rat GRF as that of a 43-residue peptide with a free carboxy-terminus. We now have isolated and characterized hypothalamic GRF from porcine, bovine, ovine and caprine origins (Böhlen *et al.*, 1983*b*; Brazeau *et al.*, 1984; Esch *et al.*, 1983). They are all 44-residue peptides with an amidated carboxyl terminus, closely related to the sequence of the human tumor-derived GRF.

We have also sequenced rat hypothalamic GRF and have confirmed the primary structure reported by Spiess *et al.* (1983) as that of a 43 amino acid residue peptide with a free acid carboxyterminal (Böhlen *et al.*, 1984).

Human hypothalamic GRF was isolated by us and shown to have the sequence of hpGRF-44-NH_2 (Ling *et al.*, 1984). The complete structure of the precursor molecule of human GRF was obtained in the Spring of 1983 by molecular cloning (Gubler *et al.*, 1983); it confirmed the complete sequence 1–44, with evidence of a Leu^{44}-Gly-Arg sequence (in the precursor) recognized as the substrate for amidation of the C-terminal residue (Leu^{44}).

Clinical studies were initiated as soon as we got clearance from the FDA early in January 1983. At the writing of this manuscript several thousand individuals have received GRF, synthesized in our laboratories by Nicholas Ling. Judging from the literature, a large number of other subjects have received synthetic GRF of various sources, as well as the fragment GRF-40. The pharmaceutical industry has already announced large-scale production of GRF (*Le Monde*, 21 September 1983). Somatocrinin (a word I have proposed to replace the initials GRF, and an obvious mirror image of the name of its physiological counterpart, somatostatin) had as spectacular a beginning as its earlier search had been slow and elusive. The remarkable series of events I have recounted was made possible only by the generosity and scientific awareness of a keen physician, passing to a group of well-prepared laboratory colleagues the rare pathology specimen that could not have been exploited otherwise.

Isolation of tumor-derived and hypothalamic GRFs

The existence of peptides with the biological activity of GRF, produced ectopically in rare cases of pancreatic islet tumors and variously located carcinoids, is well documented (Frohman *et al.*, 1980; Beck *et al.*, 1973; Caplan *et al.*, 1978; UzZafar *et al.*, 1979; Leveston *et al.*, 1981; Thorner *et al.*, 1982; Böhlen *et al.*, 1982). Several GRF peptides have now been isolated from two separate pancreatic tumors that caused acromegaly. From one tumor (tumor 1) provided by G.

Fig. 2. Isolation of GRF peptides from a pancreatic islet cell tumor* (provided by G. Sassolas, Lyons, France). **a (top):** Gel filtration of acidic tissue extract (50 ml) of 7 g tumor on Sephadex G-75 (120 × 4.5 cm). Eluent 5 M acetic acid, flow rate 60 ml/h, fraction volume 15 ml. The first and last absorbance peaks correspond to exclusion and salt volumes of the column respectively. **b (center):** Semi-preparative reverse-phase HPLC of pool of bioactive gel filtration fractions 43-58 on C18 column using pyridine formate/n-propanol as mobile phase. Flow rate 0.8 ml/min. Bioassay was performed with pooled or individual aliquots (5-10 μl) of column fractions. To avoid losses due to lyophilization, pooled gel filtration fractions were applied directly to the reverse-phase column by pumping the sample onto the column prior to starting the elution gradient. **c (bottom):** Reverse-phase HPLC of bioactive fractions from semi-preparative chromatograph (**b**) using analytical C18 column in conjunction with 0.2% (v/v) heptafluorobutyric acid/acetonitrile. Flow rate 0.6 ml/min. The sample was loaded (after 3-fold dilution) as described above. AUFS: absorbance units full scale. Bioassay was performed with aliquots of individual or pooled column fractions that had been dried in the presence of 100 μg serum albumin.

*This tumor also contained several forms of immunoreactive somatostatin (panel A), two of which were isolated and characterized as somatostatin-14 (fractions 78-88) and somatostatin-28 (fractions 52-65) by amino acid sequence.

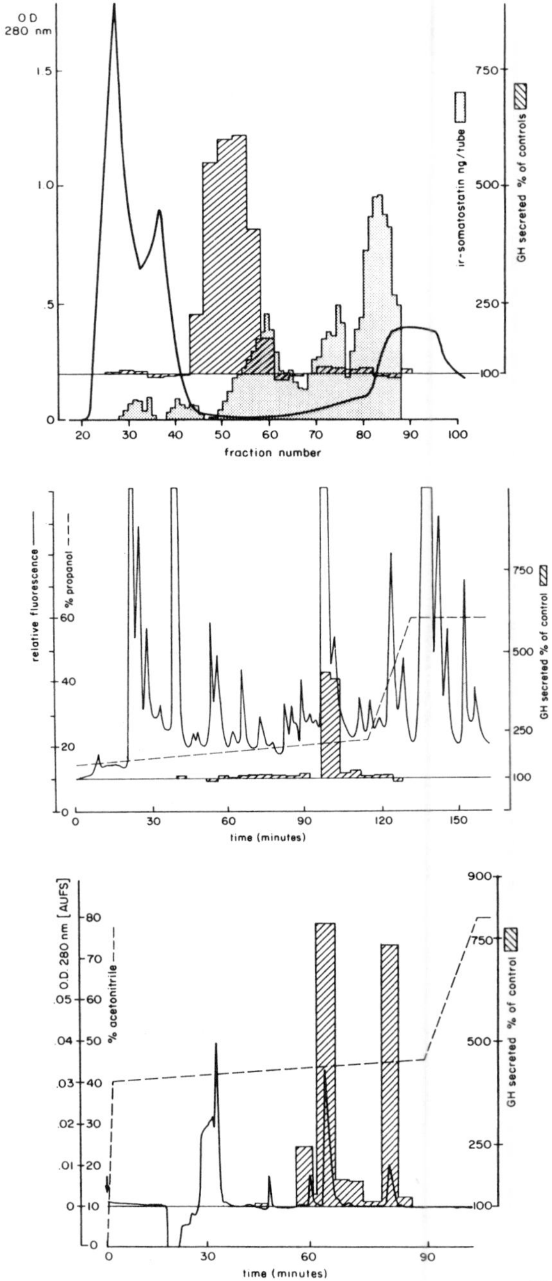
O D
280 nm
ir-somatostatin ng/tube
GH secreted % of controls
fraction number
relative fluorescence
% propanol
GH secreted % of control
time (minutes)
O.D 280 nm [AUFS]
% acetonitrile
GH secreted % of control
time (minutes)

Sassolas (Sassolas *et al.*, 1983) three GRF peptides were first isolated by our group and their structures determined: hpGRF-44, hpGRF-40 and hpGRF-37, consisting of 44, 40 and 37 amino acids respectively (Böhlen *et al.*, 1983*a*). Isolation was accomplished by means of a rapid and efficient 4-step procedure, including (a) tissue extraction in 0.3 M hydrochloric acid, (b) gel filtration and (c,d) two steps of reverse-phase HPLC using systems of different solute selectivities (Fig. 2). Peptide purification was monitored with a highly sensitive *in vitro* bioassay that tests the ability of column fractions to stimulate the release of growth hormone from rat anterior pituitary cells in serum-free monolayer culture (Brazeau *et al.*, 1982*a*). The second tumor (tumor II), provided by M. Thorner, University of Virginia (Thorner *et al.*, 1982) contained only one GRF form, identical to hpGRF-40 isolated from tumor I. It was isolated independently by our group (Böhlen *et al.*, 1982; Esch *et al.*, 1982), using the methodology outlined above, and by Rivier *et al.* (1982), using a similar microanalytical approach.

The isolation of GRFs from these two pancreatic tumors reveals a considerable degree of dissimilarity among GRF-producing neoplastic tissues. Tumor I contained three forms of GRF, with the 44-residue COOH-terminal amidated peptide (the 'mature' peptide) possessing the highest potency in our *in vitro* assay. The carboxyl terminally-shortened peptides hpGRF-40 and hpGRF-37, with a free COOH-terminal, which possess reduced *in vitro* bioactivity, may represent products arising from multiple parallel processing of the GRF precursor protein, or alternatively may be (still active) proteolytic degradation products of hpGRF-44. The contrasting finding that tumor II only contains hpGRF-40 but not the parent peptide hpGRF-44 suggests that pro-GRF is processed differently in different tumor tissues. On the other hand the precursor for GRF has recently been cloned from both tumors (Gubler *et al.*, 1983; Mayo *et al.*, 1983) and both show to the full 1–44 with the classic amidation signal -(Leu)-Gly-Arg- for the carboxyterminal. This view is also supported by our finding of two major forms of GRF in a human lung carcinoid (Böhlen *et al.*, 1982). The two islet cell tumors differed also with respect to their GRF concentration. While tumor I yielded 7.5 nmol total GRF/g of tissue, only 0.45 nmol GRF/g was isolated from tumor II. Finally, dissimilarity exists with regard to expression of the somatostatin gene. Tumor I, but not tumor II, was found to contain significant quantities of both somatostatin-14 and somatostatin-28 (Böhlen *et al.*, 1983*a*).

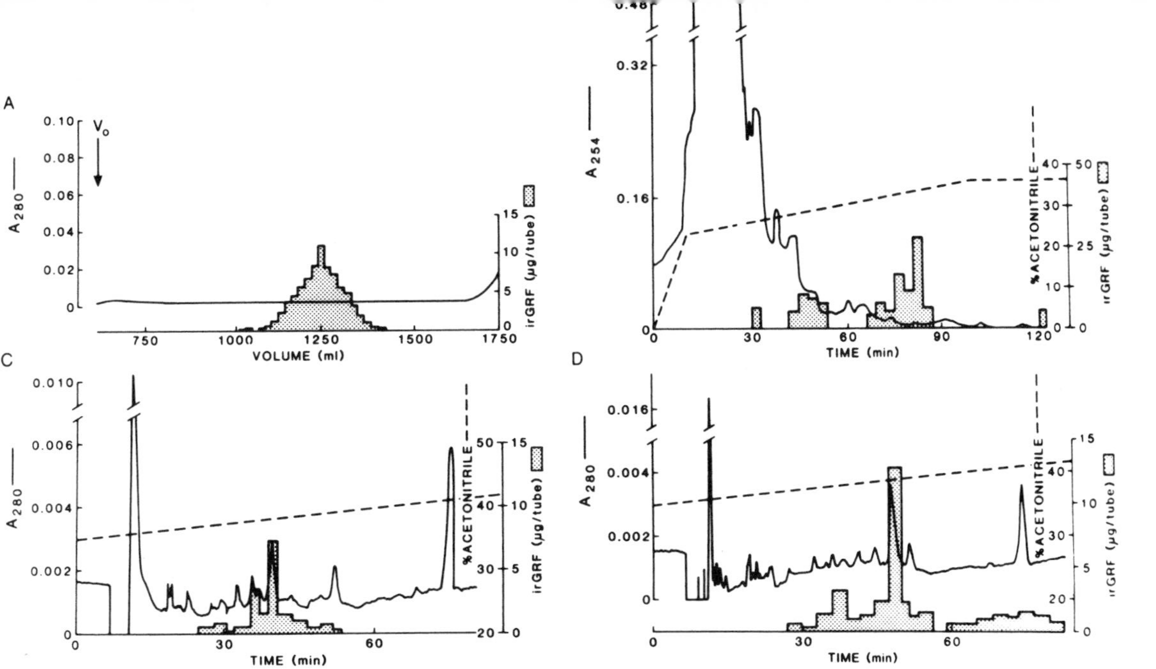

Fig. 3. Isolation of GRF from a batch of 1,032 human pituitary stalk-hypothalamic median eminence fragments. **a:** The immunoaffinity purified irGRF was chromatographed on a Sephadex G-75 column (4.5 × 117 cm, V_{bed} = 1,900 ml) and developed in 1 M acetic acid/0.2% 2-mercaptoethanol (v/v) at 1 ml/min. **b:** The gel filtered irGRF fractions eluting between 1,098 and 1,368 ml were pooled and pumped onto a semipreparative 5 μm particle size, 1 × 25 cm Ultrasphere ODS column (Altex) utilizing a 0.25 M triethylammonium phosphate, pH 3.0/acetonitrile mobile phase. Fractions of 2.5 ml were collected at 1 ml/min. **c:** The irGRF species eluting between 78 and 87 min. in chromatogram B were pooled and purified on a 7 μm particle size, 0.46 × 25 cm Aquapore RP-300 column (Brownlee Labs.) using a 0.2% (v/v) heptafluorobutyric acid/acetonitrile solvent system. Fractions of 2.5 ml were collected at 1 ml/min. **d;** The irGRF species eluting between 72 and 77 min in chromatogram **b** were pooled and purified in the same manner as in chromatogram **c**.

At all chromatography steps 1% aliquots of the column fractions were subjected to radioimmunoassay after drying in a vacuum centrifuge (Savant) in the presence of 100 μg serum albumin.

Isolation of hypothalamic GRFs

Is hpGRF identical in structure to the physiological growth hormone-releasing factor in the human hypothalamus? In an attempt to answer this question, we obtained several thousand human hypothalamic fragments and processed them by gel filtration and reverse-phase HPLC, as in the above methodology for the pancreatic tumors, with an additional step (immediately preceding G75 gel filtration) of immunoaffinity chromatography using Affi-gel-coupled polyclonal antibodies raised against hpGRF-40. Following the GRF molecule(s) by bioassay and two radioimmunoassays of different epitope specificity revealed the presence of two major forms of ir-GRF which coelute with human pancreas GRFs, hpGRF-44-NH_2 and hpGRF-40 under highly resolutive conditions (Ling *et al.*, 1984) (Fig. 3). The bioactive material coeluting with hpGRF-44-NH_2 is recognized by two antibodies which

Table 1 *Amino acid composition of the purified irGRF from Figs. 3c and 3d*

Amino Acid	irGRF peak eluting at 40 min. from Fig. 3c	ir GRF peak eluting at 49 min. from Fig. 3d
Asx	3.59 ± 0.05 (4)*	3.72 ± 0.11 (4)**
Thr	0.99 ± 0.06 (1)	0.97 ± 0.02 (1)
Ser	3.76 ± 0.10 (4)	3.84 ± 0.14 (4)
Glx	6.74 ± 0.06 (7)	6.63 ± 0.11 (7)
Gly	3.93 ± 0.11 (3)	3.71 ± 0.11 (3)
Ala	4.07 ± 0.16 (4)	4.84 ± 0.05 (5)
Val	0.86 ± 0.01 (1)	1.03 ± 0.01 (1)
Met	0.89 ± 0.21 (1)	0.81 ± 0.10 (1)
Ile	1.71 ± 0.02 (2)	1.94 ± 0.22 (2)
Leu	3.59 ± 0.20 (4)	4.84 ± 0.33 (5)
Tyr	2.18 ± 0.12 (2)	2.12 ± 0.09 (2)
Phe	0.90 ± 0.04 (1)	1.08 ± 0.04 (1)
His	0.00	0.00
Trp	0.00	0.00
Lys	2.27 ± 0.15 (2)	2.28 ± 0.08 (2)
Arg	4.50 ± 0.04 (4)	6.19 ± 0.01 (6)
Cys	0.00	0.00
Pro	0.00	0.00

Ten to 15 pmol of peptide were hydrolyzed at 110°C for 20 hours in 50 μl constant-boiling HCl containing 2% (v/v) thioglycollic acid and analyzed in a Liquimat III amino acid analyzer (Kontron) as described (7). Values are mean ± standard deviation from two determinations and are not corrected for hydrolysis losses.

* Values in parentheses correspond to the integer amino acid composition of hpGRF-40.

** Values in parentheses correspond to the integer amino acid composition of hpGRF-44.

Table 2 *Gas-phase sequence analysis of hGRF-44*

Cycle No. N	Residue No.	> PhNCS-AA	Yield (pmol)
1	1	Tyr	149
2	2	Ala	107
3	3	Asp	43.8
4	4	Ala	136
5	5	Ile	142
6	6	Phe	92.5
7	7	Thr	54.0
8	8	Asn	98.2
9	9	Ser	62.2
10	10	Tyr	88.2
11	11	Arg	69.5
12	12	Lys	77.0
13	13	Val	86.1
14	14	Leu	108
15	15	Gly	62.3
16	16	Gln	22.8
17	17	Leu	53.8
18	18	Ser	31.0
19	19	Ala	50.6
20	20	Arg	33.2
21	21	Lys	20.8
22	22	Leu	37.1
23	23	Leu	7.87
24	24	Gln	21.4
25	25	Asp	8.26
26	26	Ile	22.6
27	27	Met	19.5
28	28	Ser	15.5
29	29	Arg	9.27
30	30	Gln	7.63
31	31	Gln	7.18
32	32	Gly	12.9
33	33	Glu	7.65
34	34	Ser	2.59
35	35	Asn	6.16
36	36	Gln	0.95
37	37	Glu	8.36
38	38	Arg	5.68
39	39	Gly	3.62
40	40	Ala	5.40
41	41	Arg	3.86
42	42	Ala	3.72
43	43	Arg	0.97
44[a]	44	X	—

The amount applied was 500 pmol, the average repetitive yield was 90.9%, and the initial yield was 40.0%. > PhNCS, phenylthiohydantoin; AA, amino acid.

The carboxyl-terminal leucine was identified by comparison of the above sequence for hGRF-44(1-43) with the amino acid composition for hGRF-44.

are directed against the amidated carboxyl-terminal sequence and the central portion of the GRF-44 peptide. The bioactive GRF which coelutes with hpGRF-40 reacts only with the antibody recognizing the central portion of hpGRF (Fig. 3). Microsequencing of the material eluting as hpGRF-44-NH_2 led to establishing the full primary structure of human hypothalamic GRF to be identical to that of hpGRF-44-NH_2 (Tables 1 and 2).

Several groups have, in the past, reported or claimed purification and/or characterization of hypothalamic GRF from various species (Dhariwal *et al.*, 1965; Schally *et al.*, 1969, 1971; Veber *et al.*, 1971; Wilber *et al.*, 1971; Malacara *et al.*, 1972; Stachura *et al.*, 1972; Wilson *et al.*, 1974; Johansson *et al.*, 1974; Boyd *et al.*, 1978; Nair *et al.*, 1978; Brazeau *et al.*, 1981*b*; Arimura *et al.*, 1983; Sykes and Lowry, 1983). None of these attempts resulted in the determination of the chemical structure of an unquestionable hypothalamic growth hormone-releasing factor. This was due primarily to the non-specificity of the bioassays used to follow the presumed GRF in the purification procedures. It was not until 1983 that several hypothalamic GRFs from various animal species were isolated, chemically characterized, and shown to have true GH-releasing activity using acceptable criteria *in vitro* and *vivo*. By the batch-processing of 80,000 rat hypothalami through acid extraction, gel filtration and several steps of preparative, semi-preparative and analytical reverse-phase HPLC, Spiess *et al.* (1983) isolated 3.4 nmol of a GRF characterized as a 43-amino acid peptide with a free C-terminus. This structure shows major homology with human GRF, with 15 substitutions or deletions from that of human GRF-44. Recently our group isolated and characterized porcine hypothalamic GRF (Böhlen *et al.*, 1983*b*) (Table 3). The antibody raised against hpGRF-40 which recognizes the central portion of the GRF sequence was found to bind with porcine GRF to the same

Table 3 *Amino acid sequences of mammalian hypothalamic GRFs. The letters h, p, c, b, o and r designate the human, porcine, caprine, bovine, ovine and rat species respectively*

	5 10 15 20 25 30 35 40	
hGRF	YADAIFTNSYRKVLGQLSARKLLQDIMSRQQGESNQERGARARL	-NH_2
pGRF	YADAIFTNSYRKVLGQLSARKLLQDIMSRQQGERNQEQGARVRL	-NH_2
c, bGRF	YADAIFTNSYRKVLGQLSARKLLQDIMNRQQGERNQEQGAKVRL	-NH_2
oGRF	YADAIFTNSYRKILGQLSARKLLQDIMNRQQGERNQEQGAKVRL	-NH_2
rGRF	HADAIFTSSYRRILGQLYARKLLHEIMNRQQGERNQEQRSRFN	-OH

```
                                                                         50
Clone  8:  TGGGAACGCCAGGCGGCTGCCAGAGCAAACACCCAGCCCAGGGCCCCTGGATTTGAGCAGTGCCT
Clone 21:                                                   AGGGCCCCTGGATTTGAGCAGTGCCT

                                         100
CGGAGCAGAGGGATATCTGCCGCATCAGGTGCCACCCCGGGTGAAGG ATG CCA CTC TGG GTG TTC
CGGAGCAGAGGGATATCTGCCGCATCAGGTGCCACCCCGGGTGAAGG ATG CCA CTC TGG GTG TTC
           preproGRF-108                        Met Pro Leu Trp Val Phe
           preproGRF-107                        Met Pro Leu Trp Val Phe

                          150
TTC TTT GTG ATC CTC ACC CTC AGC AAC AGC TCC CAC TGC TCC CCA CCT CCC CCT
TTC TTT GTG ATC CTC ACC CTC AGC AAC AGC TCC CAC TGC TCC CCA CCT CCC CCT
Phe Phe Val Ile Leu Thr Leu Ser Asn Ser Ser His Cys Ser Pro Pro Pro Pro
Phe Phe Val Ile Leu Thr Leu Ser Asn Ser Ser His Cys Ser Pro Pro Pro Pro
            10                                       20
                        200
TTG ACC CTC AGG ATG CGG CGG TAT GCA GAT GCC ATC TTC ACC AAC AGC TAC CGG
TTG ACC CTC AGG ATG CGG CGG TAT GCA GAT GCC ATC TTC ACC AAC AGC TAC CGG
Leu Thr Leu Arg Met Arg Arg Tyr Ala Asp Ala Ile Phe Thr Asn Ser Tyr Arg
Leu Thr Leu Arg Met Arg Arg Tyr Ala Asp Ala Ile Phe Thr Asn Ser Tyr Arg
                    30                                           40
            250
AAG GTG CTG GGC CAG CTG TCC GCC CGC AAG CTG CTC CAG GAC ATC ATG AGC AGG
AAG GTG CTG GGC CAG CTG TCC GCC CGC AAG CTG CTC CAG GAC ATC ATG AGC AGG
Lys Val Leu Gly Gln Leu Ser Ala Arg Lys Leu Leu Gln Asp Ile Met Ser Arg
Lys Val Leu Gly Gln Leu Ser Ala Arg Lys Leu Leu Gln Asp Ile Met Ser Arg
                                 50                                  60
        300
CAG CAG GGA GAG AGC AAC CAA GAG CGA GGA GCA AGG GCA CGG CTT GGT CGT CAG
CAG CAG GGA GAG AGC AAC CAA GAG CGA GGA GCA AGG GCA CGG CTT GGT CGT CAG
Gln Gln Gly Glu Ser Asn Gln Glu Arg Gly Ala Arg Ala Arg Leu Gly Arg Gln
Gln Gln Gly Glu Ser Asn Gln Glu Arg Gly Ala Arg Ala Arg Leu Gly Arg Gln
                                     70                       *
    350                                                                400
GTA GAC AGC ATG TGG GCA GAA CAA AAG CAA ATG GAA TTG GAG AGC ATC CTG GTG
GTA GAC AGC ATG TGG GCA GAA CAA AAG CAA ATG GAA TTG GAG AGC ATC CTG GTG
Val Asp Ser Met Trp Ala Glu Gln Lys Gln Met Glu Leu Glu Ser Ile Leu Val
Val Asp Ser Met Trp Ala Glu Gln Lys Gln Met Glu Leu Glu Ser Ile Leu Val
    80                                      90
                                                                450
GCC CTG CTG CAG AAG CAC AGC AGG AAC TCC CAG GGATGAAGATTCCTCCTGTGACCCGGG
GCC CTG CTG CAG AAG CAC     AGG AAC TCC CAG GGATGAAGATTCCTCCTGTGACCCGGG
Ala Leu Leu Gln Lys His Ser Arg Asn Ser Gln Gly
Ala Leu Leu Gln Lys His  ↑  Arg Asn Ser Gln Gly
            100
                                        500
CTACCTGTAGCCAAAATGCAACTGGATCCAGTTAATCCTCTCATTTCTGACCCACTTTTTCCTTTGAAAAT
CTACCTGTAGCCAAAATGCAACTGGATCCAGTTAATCCTCTCATTTCTGACCCACTTTTTCCTTTGAAAAT

                550                 570
ACAATAAAATTCCCCCATACCGGTGTGCATTTAAATGTTAAAAAAAAAAAAAAA
ACAATAAAATTCCCCCATACCGGTGTGCATTTAAA  ↑  AAAAAAAAAAAAAAA
```

Fig. 4. Nucleotide sequences of the cDNAs correspond to two hpGRF-44 mRNAs and deduced amino acid sequences of pre-pro-GRF-107 and -108. The presumed polyadenylation sequence A-A-T-A-A-A in the 3′ untranslated region is underlined. Nucleotides 419-421 and 567-570 (↑) are found in clone 8 but not in clone 21. The sequence of hpGRF-44 extends from amino acid residue 32 to amino acid residue 75 and is boxed. Basic amino acids that serve as processing signals to liberate hpGRF-44 from pre-pro-GRF are double-toxed. The putative signal peptide is indicated with a dashed line and the Gly-Arg amidation signal (amino acid residues 76 and 77) is indicated with an asterisk (*).

degree as hpGRF-44 or hpGRF-40 (parallel displacement curves in the RIA). The RIA established for human GRF was therefore used for monitoring peptide purification. A carboxyl terminal amidated 44-residue peptide with only 3 substitutions from the sequence of human GRF (see Table 3) was isolated from 2,500 porcine hypothalami, using an approach that included immunoaffinity chromatography besides gel filtration and two reverse-phase HPLC steps. Immunoaffinity chromatography with Affi-gel-bound purified IgG antibodies, as designed by Nicholas Ling, provided a powerful tool for the purification of pGRF from crude tissue extracts. Owing to the high efficiency of this step it was possible to isolate quantities of GRF sufficient for full structural characterization from relatively small amounts of tissue (6 nmol pGRF-44 from 2,500 hypothalami). Other isolation procedures which omit the immunoaffinity step yielded pure pGRF as well but with substantially lower yield.

Using the same purification scheme, including the immunoaffinity chromatography step, we have also isolated and characterized bovine hypothalamic GRF (see Table 3). We have also isolated hypothalamic ovine GRF and hypothalamic caprine (goat) GRF, again using the purification scheme described above. The primary structures of these GRFs have all been established (Table 3). They are all 44-residue carboxyterminal amidated peptides. Only rat hypothalamic GRF, isolated and sequenced by Spiess *et al.* (1983) and by us (Böhlen *et al.*, 1984), has a free carboxyl terminal and is composed of 43 amino acid residues (Table 3). All these structures share considerable homologies, mostly in their amino-terminal end.

Structure of pre-pro-GRF

The primary structures of two precursors for human pancreas GRF were established by molecular cloning and DNA sequence analysis of cDNA coding for pre-pro-GRF-mRNA (Gubler *et al.*, 1983) (Fig. 4). The two forms of mRNA code for pre-pro-GRF-107 and -108. The two polypeptides contain the sequence of hpGRF-44 flanked by basic processing sites (Fig. 5). The precursors include a putative signal sequence and a COOH-terminal amidation signal for hpGRF-44, the latter in keeping with our demonstration that the carboxyl terminus of hpGRF-44 is amidated. Pre-pro-GRF-108 differs from pre-pro-GRF-107 by the insertion of serine in the COOH-terminal portion of the precursor (in position 104). The molecular weight of pre-pro-GRF as determined by *in vitro* translation of tumor poly(A^+)-

Fig. 5. Schematic representation of pre-pro-GRF-107 amd -108. The sequence of hpGRF-44 (in black) is flanked by processing sites consisting of basic amino acids. A glycine residue immediately adjacent to the carboxyl terminus of GRF mediates the amidation of hpGRF-44. A hydrophobic sequence at the amino terminus (probably 20 amino acids in length) represents the signal sequence. Two cryptic sequences of 9 (amino terminal) and 31 (carboxyl terminal) amino acids comprise the rest of the precursor protein. Pre-pro-GRF-107 lacks the residue serine-103.

RNA, followed by immunoprecipitation with hpGRF-specific antiserum and $NaDodSO_4$ polyacrylamide gel electrophoresis, was shown to be approximately 13,000, which agrees with the size of a 107 or 108 amino acid peptide (Fig. 4). Mayo *et al.* (1983) have confirmed these results in cloning the precursor of GRF from the tumor in which only GRF-40 was found and isolated.

Determination of the primary structure of tumor-derived and hypothalamic GRFs

The strategy and methodology employed in each instance are identical; thus only the structural elucidation of hpGRF-44, considered as a representative study, will be discussed in detail. It was also the original approach that led to the first characterization of the primary sequence of GRF in 1982.

Amino acid analyses. Peptides were hydrolyzed in sealed evacuated ignition tubes containing five μl 6N HCl and 7% thioglycollic acid for 24 hours at 110°C (Böhlen and Schroeder, 1982). Amino acid analyses of 10–100 pmols of peptide hydrolyzates were performed with a Liquimat III amino acid analyzer (Kontron, Zurich, Switzerland) equipped with a proline conversion fluorescence detection system (Böhlen and Mellet, 1979).

Edman degradation. The sequential degradation of peptides was performed with an Applied Biosystems Model 470A gas-phase sequencer as described (Hunkapiller *et al.*, 1983) with several modifications. The dipeptide, Phe-Leu, was used instead of Gly-Gly to precycle the Polybrene (Aldrich). Longer coupling (1200 sec.) times in all cycles and a longer cleavage time (1665 sec.) in the first cycle were used to enhance the repetitive and initial yields, respectively. Finally,

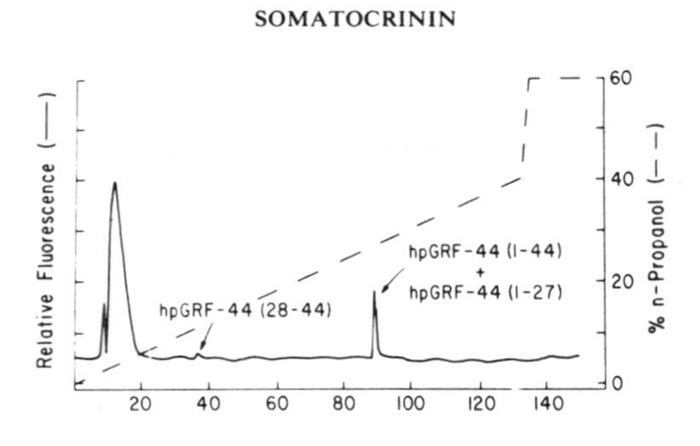

Fig. 6. Reverse phase liquid chromatography of a cyanogen bromide digest of hpGRF-44. Digestion and chromatography protocols were as described in Methods. Gradient elutions were at 0.6 ml/min. and room temperature with pyridine formate/n-propanol. The column effluents were collected in 1.1 ml fractions and monitored with an automatic fluorescamine detection system (Böhlen *et al.*, 1975).

an extensive methanol wash (180 sec., 3.8 ml) was used to clean the conversion flask after each cycle and thus reduce cycle-to-cycle carry over. The phenylthiohydantoin amino acid derivatives were unambiguously identified by high pressure liquid chromatography as described (Hunkapiller and Hood, 1983).

Cyanogen bromide digestions. Six nmols of peptide in 1 ml of [0.5% heptafluorobutyric acid, 48% acetonitrile] were rapidly dried in a 17×100 mm polypropylene tube using a Savant vacuum centrifuge. Upon dissolution of the peptide with 100 μl 70% formic acid, a small crystal of cyanogen bromide was added and the reaction mixture capped and vortexed. After incubation for 12 hours in the dark, the cyanogen bromide-generated peptide fragments were purified with an Altex Model 322 high pressure liquid chromatography system equipped with a Brownlee RP-18 guard column (10 μm; 4.6×30 mm), a Brownlee RP-18 analytical column (5 μm; 4.6×250 mm) and a modified pyridine formate/n-propanol system (Rubenstein *et al.*, 1977) consisting of 1% pyridine, 1.5% formate in solvents A and B, with B also containing 60% n-propanol. The isolation of a homogeneous preparation of the cyanogen bromide digestion fragment, hpGRF-44(1-27), required rechromatography of the material in the peak (Fig. 6) containing this fragment and the undigested hpGRF-44.

High pressure liquid chromatography comparison of native hpGRF-44 and synthetic replicates

The synthetic replicates of hpGRF-44 were synthesized in two forms, i.e., with a free acid and an amidated carboxyl terminus. Two different

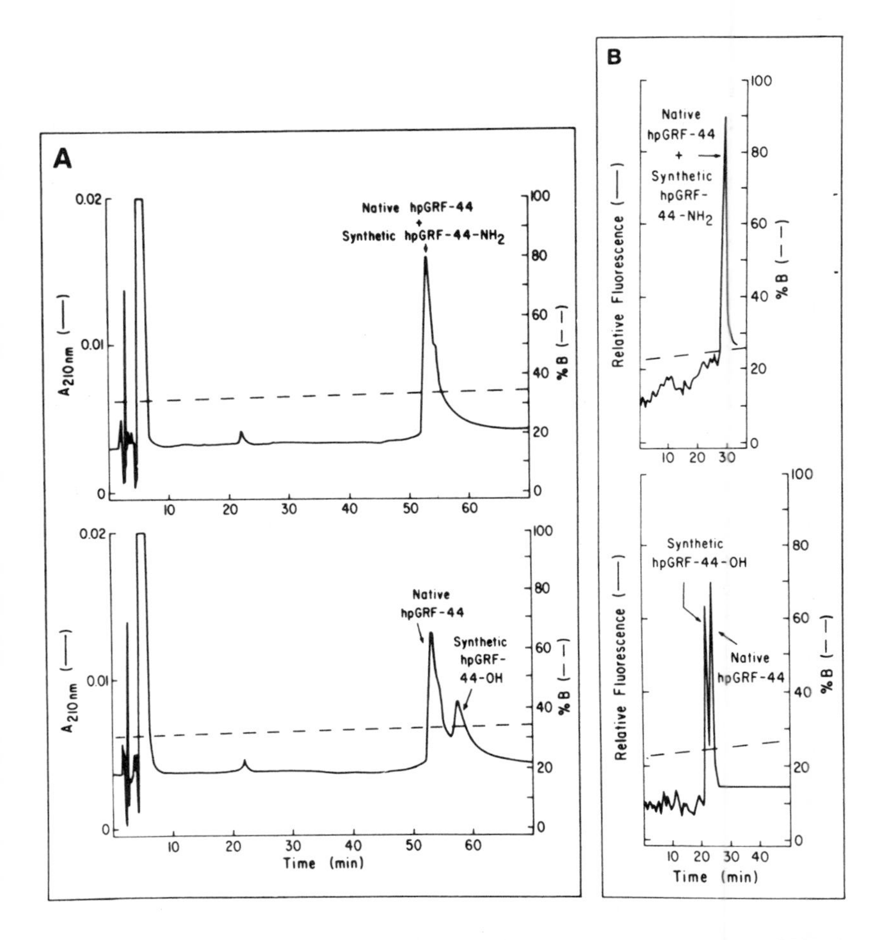

Fig. 7. Reverse phase and ion exchange liquid chromatography of native hpGRF-44 with its synthetic replicates containing either a free acid or an amidated carboxy terminus. 250–500 pmol quantities of each peptide were chromatographed at room temperature.

a: Reverse-phase gradient elutions of the peptide at 1.5 ml/min. employed an 80-min. linear gradient from 31–35% B, where solvent A was 0.25 N triethylammonium phosphate, pH 3.0 and solvent B was 20% 0.25 N triethylammonium phosphate, pH 3.0, and 80% acetonitrile.

b: Ion exchange gradient elutions of the peptide at 0.6 ml/min. employed a 60-min. linear gradient from 22.5 to 27.5% B, where solvent A was 0.045 N sodium acetate, pH 6.01 and solvent B was 0.91 N sodium acetate, pH 6.01. Additionally, both solvents contained 30% n-propanol.

high pressure liquid chromatography systems were employed to effect separation between the free acid and amidated forms of synthetic replicates and, in turn, to ascertain with which form the native peptide would co-elute. An Altex Model 322 high pressure liquid chromatography system with a Brownlee RP-18 (10 μm; 4.6 × 30 mm) guard column, an Altex Ultrasphere ODS (5 μm; 4.6 × 250 mm) analytical column and triethylammonium phosphate/acetonitrile buffers (Rivier, 1978) were used to separate the free acid and amidated carboxy terminal forms of the peptide by reverse-phase methodology. Chromatography conditions are described in Fig. 7.

An Altex Model 332 high pressure liquid chromatography system with a TSK IEX-535K (10 μm; 4.6 × 150 mm) carboxymethyl ion-exchange analytical column, 0.045 M and 0.91 M sodium acetate, pH 6.0 buffers, each containing 30% n-propanol, were used in the second system to achieve a normal-phase separation of the peptide derivatives. Chromatography conditions are described in Fig. 7.

The primary structure of hpGRF-44 was established by sequence analyses of the intact peptides and their cyanogen bromide digestion fragments. The structure of hpGRF-44(1-40) was established by Edman degradation of the intact peptide. Confirmation of these data

Table 4 *Amino acid compositions of hpGRF-44 and its cyanogen bromide digestion fragments*

Amino Acid	hpGRF-44 (n = 8)	hpGRF-44(1-27) n = 1)	hpGRF-44(28-44) (n = 2)
Asx	4.05 ± 0.27 (4)*	2.85 (3)*	1.02 (1)*
Thr	0.93 ± 0.09 (1)	0.75 (1)	0
Ser	3.76 ± 0.20 (4)	1.95 (2)	1.91 (2)
Glx + Hse	6.87 ± 0.29 (7)	2.90 (3)	4.71 (5)
Gly	3.43 ± 0.48 (3)	1.30 (1)	2.37 (2)
Ala	4.93 ± 0.19 (5)	3.15 (3)	1.94 (2)
Val	0.89 ± 0.06 (1)	0.87 (1)	0
Met	1.04 ± 0.16 (1)	0	0
Ile	1.84 ± 0.15 (2)	1.67 (2)	0
Leu	4.99 ± 0.22 (5)	4.00 (4)	1.05 (1)
Tyr	2.10 ± 0.36 (2)	2.15 (2)	0
Phe	0.96 ± 0.37 (1)	1.20 (1)	0
His	0.00	0	0
Trp	0.00	0	0
Lys	2.02 ± 0.37 (2)	2.37 (2)	0
Arg	6.09 ± 0.61 (6)	2.35 (2)	4.00 (4)
Cya	0.00	0	0
Pro	0.00	0	0

* Values in parentheses were deduced from sequence analyses

Table 5 *Sequence analysis of hpGRF-44*
Amount applied: 1.500 pmol
Initial Yield: 37.3%   Average Repetitive Yield: 89.7%

Cycle No. (N)	Residue No.	>PhNCS-AA	Yield (pmol)	Carry over from (N-1) (pmol)
1	1	Tyr	719	—
2	2	Ala	531	29.0
3	3	Asp	281	21.4
4	4	Ala	551	19.7
5	5	Ile	322	21.3
6	6	Phe	341	14.8
7	7	Thr	86.3	11.9
8	8	Asn	237	0
9	9	Ser	44.2	15.0
10	10	Tyr	200	2.7
11	11	Arg	163	20.2
12	12	Lys	178	30.6
13	13	Val	200	20.4
14	14	Leu	153	26.5
15	15	Gly	89.3	28.4
16	16	Gln	106	43.0
17	17	Leu	105	57.3
18	18	Ser	8.9	70.8
19	19	Ala	84.7	3.0
20	20	Arg	50.0	37.3
21	21	Lys	54.4	41.9
22	22	Leu	91.0	28.1
23	23	Leu	73.0	—
24	24	Gln	33.8	32.1
25	25	Asp	28.7	29.7
26	26	Ile	21.5	10.6
27	27	Met	34.6	22.1
28	28	Ser	5.3	4.0
29	29	Arg	20.9	2.7
30	30	Gln	17.1	12.9
31	31	Gln	58.4	—
32	32	Gly	22.0	15.7
33	33	Glu	17.2	12.5
34	34	Ser	3.4	12.8
35	35	Asn	14.0	0.8
36	36	Gln	8.4	8.4
37	37	Glu	11.7	0
38	38	Arg	5.3	4.1
39	39	Gly	13.0	4.0
40	40	Ala	7.5	9.2
41	41	X	—	—
42	42	X	—	—
43	43	X	—	—
44	44	X	—	—

In all degradations the initial yield estimates were obtained by extrapolation of all phenylthiohydantoin amino acid yields back to cycle number one using the average repetitive yield.

and determination of the remaining carboxy terminal sequence were accomplished by cyanogen bromide digestion of the hpGR-44, high pressure liquid chromatographic isolation of the digestion fragments (Fig. 6) and subsequent structural characterization of these fragments by amino acid analyses (Table 4) and Edman degradations (Tables 5, 6 and 7). These results were obtained with a total of 7.5 nmol peptide and are summarized in Table 8.

Table 6 *Sequence analysis of the CNBr fragment: hpGRF-44(1-27)*
Amount applied: 540 pmol
Initial Yield: 57.4% Average Repetitive Yield: 88.9%

Cycle No. (N)	Residue No.	>PhNCS-AA	Yield (pmol)	Carry over from (N-1) (pmol)
1	1	Tyr	221	—
2	2	Ala	239	7.4
3	3	Asp	150	11.9
4	4	Ala	219	17.3
5	5	Ile	194	28.3
6	6	Phe	178	18.8
7	7	Thr	62.4	11.0
8	8	Asn	176	10.4
9	9	Ser	33.1	14.2
10	10	Tyr	106	0
11	11	Arg	158	20.6
12	12	Lys	217	0
13	13	Val	77.8	22.5
14	14	Leu	92.3	15.1
15	15	Gly	32.9	14.4
16	16	Gln	49.9	13.7
17	17	Leu	54.7	9.7
18	18	Ser	5.8	14.2
19	19	Ala	25.9	0
20	20	Arg	34.3	2.5
21	21	Lys	29.6	0
22	22	Leu	26.0	13.7
23	23	Leu	27.3	—
24	24	Gln	27.0	10.8
25	25	Asp	17.4	19.9
26	26	X	—	5.9
27	27	X	—	—

Table 7 *Sequence analysis of the CNBr fragment: hpGRF-44(28–44)*
Amount Applied: 490 pmol
Initial Yield: 64.3% Average Repetitive Yield: 89.1%

Cycle No. (N)	Residue No.	>PhNCS-AA	Yield (pmol)	Carry over from (N-1) (pmol)
1	28	Ser	27.6	—
2	29	Arg	274	0
3	30	Gln	188	5.4
4	31	Gln	260	—
5	32	Gly	200	0
6	33	Glu	110	37.0
7	34	Ser	30.2	20.9
8	35	Asn	127	0
9	36	Gln	114	43.1
10	37	Glu	99.7	15.2
11	38	Arg	107	22.1
12	39	Gly	115	18.5
13	40	Ala	98.6	21.8
14	41	Arg	80.6	27.3
15	42	Ala	95.1	20.1
16	43	Arg	55.4	23.4
17	44	Leu	1.5	23.5

Determination of the nature of the carboxyl terminus of hpGRF-44

Evidence for the nature of the carboxyl terminus of each of the hpGRF peptides was obtained from high pressure liquid chromatography studies in which hpGRF-44 was co-chromatographed with synthetic replicates possessing a free carboxyl or an amidated carboxyl terminus. These studies were carried out with both reverse-phase and ion-exchange high pressure liquid chromatography systems, and the results are illustrated in Fig. 7 and clearly show the peptide possessing an amidated carboxyl terminus.

At the time of this review the structures of hypothalamic growth hormone releasing peptides from porcine (Böhlen *et al.*, 1983*b*), bovine (Esch *et al.*, 1983) and rat (Spiess *et al.*, 1983) sources have also been elucidated; they are compared with the sequences of various homologous intestinal peptides in Table 9. Porcine and bovine hypothalamic GRFs are, like human GRF, 44-residue, C-terminus amidated peptides. Porcine GRF has only 3 substitutions and bovine GRF 5

Table 8 *Sequence analyses of hpGRF-44 and its cyanogen bromide fragments*

Peptide	Amount analyzed (nmol)	Primary structure
		5 10 15 20 25 30 35 40
Intact hpGRF-44	1.5	Y-A-D-A-I-F-T-N-S-Y-R-K-V-L-G-Q-L-S-A-R-K-L-L-Q-D-I-M-S-R-Q-Q-G-E-S-N-Q-E-R-G-A-X-X-X-X
CNBr fragment hpGRF–44 (1–27)	0.54	Y-A-D-A-I-F-T-N-S-Y-R-K-V-L-G-Q-L-S-A-R-K-L-L-Q-D-X-X
CNBr fragment hpGRF–44 (28–44)	0.49	S-R-Q-Q-G-E-S-N-Q-E-R-G-A-R-A-R-L

Table 9 *Sequence analyses of GRF peptides and various intestinal peptides*

	5 10 15 20 25 30 35 40
hpGRF	YADAIFTNSYRKVLGQLSARKLLQDIMSRQQGESNQERGARARL-NH_2
pGRF	YADAIFTNSYRKVLGQLSARKLLQDIMSRQQGERNQEQGARVRL-NH_2
rGRF	HADAIFTSSYRRILGQLYARKLLHEIMNRQQGERNQEQRSRFN-OH
PHI-$27_{porcine}$	HADGVFTSSYRRILGQLSAKKYLESLI-NH_2
$VIP_{porcine}$	HSDAVFTDNYTRLRKQMAVKKWLNSILN-NH_2
Glucagon	HSQGTFTSDYSKYLDSRRAQDFVQWLMNT-OH
$Secretin_{porcine}$	HSDGTFTSELSRLRDSARLKRLLQGLV-NH_2

substitutions from the amino acid sequence of human GRF. Rat GRF shows 14 amino acid substitutions from the sequence of human GRF.

Synthetic replicates of GRFs: Structure activity relationships

Total synthesis of hGRF-44, hGRF-40, hGRF-37 and fragments thereof, as well as of pGRF, bGRF, oGRF and rGRF, was achieved by solid-phase peptide synthesis methodology (Ling *et al.*, 1980). The synthetic peptides were tested in the normal rat anterior pituitary cell monolayer culture system (Brazeau *et al.*, 1982*a*). Of the hGRF peptides tested, hGRF-44 was found to be the most potent (ED_{50} = 15 pM; ED_{max} = 100 pM). Based on a potency ranking of 100 for hGRF-44, progressive deletion of the COOH-terminal residues of hGRF-44 resulted in a gradual loss of activity until hGRF-(1-28)-OH as shown in Table 10 and Table 11. Fragments shorter than (1-28) possess very little activity. These studies indicate that the biologically active core of the molecule resides in the NH_2-terminal part of the molecule. Furthermore, amidation of the COOH-terminal residue increases the potency of all the shortened hGRF peptide fragments approximately 1.5-fold over that of the free acid forms (Table 11). Met-27 is not an exclusive requirement for biological activity as it can be replaced by norvaline (see Table 12). Similarly, Tyr-1 has been replaced by other amino acids shown in Table 12. Except for His-1 substitution all other modifications yielded less potent analogs. Interestingly, rat GRF is more potent (1.5x) than hGRF-44, whereas pig GRF is ca. 75% the potency of hGRF-44, with overlapping confidence limits, and bGRF may be somewhat less potent than pGRF, all the assays being done with rat pituitary cells *in vitro*.

Table 10 *Relative potencies of COOH-terminal deletion analogs of hpGRF-44-NH_2*

Analogs	Potency	(95% confidence limits)
hpGRF-44-NH_2	100	
hpGRF(1-44)OH	61	(50–75)
hpGRF(1-40)NH_2	49	(38–62)
hpGRF(1-40)OH	30	(25–37)
hpGRF(1-37)NH_2	28	(23–33)
hpGRF(1-37)OH	12	(9–16)
hpGRF(1-34)OH	17	(12–25)
hpGRF(1-31)OH	9	(6–12)
hpGRF(1-28)OH	6	(4–9)
hpGRF(1-24)OH	0.014	—
hpGRF(1-21)OH	<0.002*	—
hpGRF(1-19)OH	<0.001*	—

* Dose-response curve not parallel to standard.

Table 11 *Relative potencies of C-terminal deletion analogs of hGRF*

Analogs	Potency	95% confidence limits
hGRF(1-44)NH_2	1	
hGRF(1-44)OH	0.70	0.56–0.87
hGRF(1-40)NH_2	0.91	0.72–1.14
hGRF(1-40)OH	0.34	0.27–0.43
hGRF(1-37)NH_2	0.50	0.40–0.62
hGRF(1-37)OH	0.27	0.20–0.36
hGRF(1-34)OH	0.23	0.19–0.30
hGRF(1-31)NH_2	0.68	0.52–0.89
hGRF(1-31)OH	0.40	0.29–0.55
hGRF(1-30)NH_2	0.51	0.39–0.64
hGRF(1-30)OH	0.27	0.18–0.49
hGRF(1-29)NH_2	0.51	0.37–0.70
hGRF(1-29)OH	0.25	0.20–0.31
hGRF(1-27)NH_2	0.12	0.09–0.17
hGRF(1-24)OH	0.0002	—
hGRF(1-23)NH_2	0.0024	0.002–0.003
hGRF(1-22)NH_2	0.00001	—
hGRF(1-21)NH_2	0.000001	—
hGRF(1-19)NH_2	Inactive up to 10^{-5} M	

Table 12 *Relative potencies of position 27 or 1 substituted analogs of hpGRF-40-OH*

Analogs	Potency	(95% confidence limits)
hpGRF-44-NH_2	100	
[Norval^{27}]hpGRF-40-OH	91	(74–112)
[Ala-Tyr^{1}]hpGRF-40-OH	4.2	(2.4–6.5)
[Arg-Tyr^{1}]hpGRF-40-OH	0.4	(0.1–1.1)
[Ac-Tyr^{1}]hpGRF-40-OH	10	(7–14)
[D-Tyr^{1}]hpGRF-40-OH	0.006*	—
[Ala^{1}]hpGRF-40-OH	0.84	(0.7–1.0)
[Phe^{1}]hpGRF-40-OH	3.1	(2.5–3.9)
[His^{1}]hpGRF-40-OH	54	(46.5–63.2)
[Trp^{1}]hpGRF-40-OH	0.006*	—
Rat GRF	93.7	(82.6–106.3)
Pig GRF	75	(69–81)

* dose-response curve not parallel to standard

In Vitro studies on the mechanism of action of GRF

In the early stages of these studies we systematically assayed in parallel the effects of a purified preparation of rat hypothalamic GRF along with the synthetic replicates of hGRF. As we became convinced, from data obtained in multiple approaches—the most important being the characterization of hypothalamic GRFs—that the tumor-derived material (hpGRF) was identical structurally and functionally to the GRF of hypothalamic origin, we stopped the practice of using a hypothalamic GRF reference standard in all experiments. Results of some of the early studies with the hypothalamic GRF reference standard will be included here since historically they were important data leading to the proposal of the identity of the hypothalamic or ectopically produced GRF.

Two types of *in vitro* pituitary cell preparations were used: the 'classic' monolayer tissue culture system using rat pituitary cells, kept in culture for 4 days, then used in short-term, 3- to 4-hour studies with one treatment or another, occasionally with shorter sampling times such as 5, 10, 15 or 30 minutes. The exact description of the technique for dissociation of the cells, plating and handling on day 4 have been described in detail earlier (Brazeau *et al.*, 1982*a*). The second type of *in vitro* pituitary cell preparation used in some experiments is that of a perifusion system, allowing continuous sampling of the pituitary effluent as well as very short stimulation pulses ($\geqslant 1$

minute). Preparation of cells for the perifusion system has also been described in detail (Brazeau *et al.*, 1982*a*).

RADIOIMMUNOASSAYS

RIAs for rat GH are conducted using Sinha's monkey-antimouse GH immune serum (Sinha *et al.*, 1972); RIAs for PRL, TSH, LH, FSH, are conducted using the antisera provided by the National Pituitary Agency NIADDK (Dr A. Parlow). RIAs for β-endorphin use the antiserum RB-100 (Guillemin *et al.*, 1977) prepared in this laboratory. RIA for cAMP uses Miles-Yeda anti-cAMP immune serum; the antiserum is reconstituted 1:15 and we omit the succinylation reaction; the trace is obtained from New England Nuclear (NEX130). Calculation of standard curves and experimental values are done with the use of the program described in (Faden *et al.*, 1980).

PEPTIDES

Reference standard preparation for rat hypothalamic GRF. This is a purified preparation of GRF from rat hypothalamic extract. Following gel filtration on Sephadex G75 in 30% acetic acid, a step that removes all somatostatin-14 and most of somatostatin-28, the zone of the effluent with GRF activity is further purified by two steps of HPLC. A GRF preparation so obtained from 2,400 rat hypothalamic fragments was aliquoted in 1.0 ml vials in tissue culture medium and kept frozen at −20°; 50 μl of this solution correspond to the ED_{50} in a complete dose-response curve; that amount of the extract is defined as 1 unit of GRF activity, and that preparation of hypothalamic GRF is referred to as GRF reference standard (Brazeau *et al.*, 1981*b*).

Synthetic hGRF and fragments. All synthetic replicates of the tumor-derived GRF (GRF-44) were prepared by solid-phase synthesis methods as routinely used in this laboratory (Ling *et al.*, 1980). When we refer to synthetic hGRF-44 we imply that the molecule is in the amidated form, as is the native material; on the other hand synthetic hGRF-40 or hGRF-37 refer to peptides in the free acid form, as are the native extracted peptides.

Somatostatin-28 and -14 were synthesized by solid-phase methods (Ling *et al.*, 1980).

Chemicals. IBMX (3-isobutyl-1-methylxanthine), 8Br.cAMP (Na salt), PGE_2, cycloheximide, $CoCl_2$ were purchased from Sigma Chemical Co. Cholera toxin, forskolin were purchased from Calbiochem-Behring Co.

Statistical analyses. Comparisons of the effects of various treatments were conducted by the multiple comparison test of Dunnett following an analysis of variance (program EXBIOL) (Sakiz, 1964). Multiple dose-response curves in the bioassays were analyzed for simultaneous fitting by the 4-parameter logistic equation of De Lean *et al.* (1978) (program ALLFIT). The same data were also studied by regression analysis and calculations of relative potencies with 95% confidence limits (program BIOPROG) (Rodbard, 1974).

SPECIFICITY OF GRF FOR THE RELEASE OF IR-GH

When tested in the *in vitro* assay described above, purified hypothalamic GRF, native tumor-derived GRF-40 or synthetic GRF-44, at doses ranging from 0.6 to 40 units or 3.1 to 400 femtomoles respectively which are known to reach E_{max} for stimulation of GH secretion, release only ir-GH; i.e. they have no effect on the secretion of ir-β-endorphin, FSH, LH, TSH or PRL (Table 13). Similar results have now been obtained with the synthetic replicates of hypothalamic rat, porcine and bovine GRFs.

DOSE-RESPONSE RELATIONSHIPS

In a large number of experiments, purified rat hypothalamic GRF, native tumor-derived GRF-37, GRF-40 and GRF-44 as well as synthetic hGRF-37, hGRF-40, and hGRF-44 all show identical dose-response curves when studied at doses ranging from 0.6 u to 40 u of GRF reference standard for the hypothalamic material and 3.1 to 400 fmol/ml for the various isolated or synthetic hGRFs. These doses are known to extend to E_{max} (see Fig. 8). Fig. 8 shows the results of one such experiment in which hypothalamic GRF, native hGRF-44, synthetic hGRF-44, synthetic hGRF-40, and synthetic hGRF-37 were assayed at multiple dose levels. Calculations (ALLFIT) of the statistical fit of these curves by the 4-parameter logistic polynomial model of De Lean *et al.* (1978) confirm that they have identical slopes (parameter b) and identical values for parameter d which represents the value of the maximal effect (E_{max}) of an agonist. In other words, in the *in vitro* assay, purified hypothalamic GRF and the tumor-derived GRFs, native or synthetic, have all identical effects and intrinsic activity; i.e. they must activate the cell machinery involved in the release of growth hormone through the same mechanism (action) and to the same maximal extent (effect). The results show also that each of the 3 forms of tumor-derived GRF has a different specific activity. If the potency of native hGRF-44 or of its synthetic replicate

Table 13 *Specificity of hypothalamic GRF, native or synthetic tumor-derived GRF to release GH, not TSH, PRL, FSH, LH, β-endorphin (βE)*[a]

Hypothalamic GRF (GRF reference std) in GRF units/ml	GH released[b] (ng/ml)	TSH released[c] (ng/ml)	PRL released[c] (ng/ml)	FSH released[c] (ng/ml)	LH released[c] (ng/ml)
0	870 ± 26	52 ± 3	384 ± 18	330 ± 19	1056 ± 102
0.63	1710 ± 60	47 ± 2	347 ± 12	291 ± 21	901 ± 42
1.25	2626 ± 24	78 ± 19	394 ± 7	315 ± 12	1012 ± 63
2.50	3923 ± 40	61 ± 11	375 ± 11	355 ± 6	1077 ± 15
5	5586 ± 52	27 ± 7	410 ± 11	276 ± 16	820 ± 90
10	6803 ± 46	40 ± 7	386 ± 20	251 ± 17	873 ± 57
20	7060 ± 75	61 ± 6	424 ± 33	283 ± 7	856 ± 52
40	7213 ± 122	63 ± 17	475 ± 10	358 ± 25	1169 ± 173
Native tumor-derived hpGRF-40 (fmol/ml)					
6.3	1903 ± 43	47 ± 6	485 ± 13	235 ± 5	761 ± 71
12.5	2163 ± 62	52 ± 13	371 ± 13	221 ± 20	679 ± 76
25	3480 ± 35	57 ± 6	388 ± 14	269 ± 11	901 ± 49
50	4820 ± 57	21 ± 4	406 ± 34	285 ± 6	947 ± 56
100	6746 ± 122	51 ± 4	369 ± 22	252 ± 24	827 ± 26
200	7070 ± 124	50 ± 3	432 ± 20	259 ± 23	844 ± 75
400	7606 ± 163	39 ± 1	405 ± 10	258 ± 4	856 ± 53

Table 13 *contd.*

Synthetic hpGRF-44-NH_2 (fmol/ml)	GH released[b] (ng/ml)	TSH released[c] (ng/ml)	PRL released[c] (ng/ml)	FSH released[c] (ng/ml)	LH released[c] (ng/ml)	βE released[2] (ng/ml)
0	342 ± 12	308 ± 80	217 ± 12	125 ± 5	569 ± 35	1157 ± 159
3.1	647 ± 9	416 ± 19	258 ± 18	130 ± 19	598 ± 42	1153 ± 66
6.3	733 ± 13	382 ± 93	248 ± 3	142 ± 14	507 ± 34	1054 ± 66
12.5	1123 ± 12	297 ± 19	305 ± 42	170 ± 15	745 ± 25	1151 ± 47
25	1447 ± 7	304 ± 65	273 ± 17	140 ± 9	592 ± 47	997 ± 80
50	1720 ± 30	343 ± 26	284 ± 9	179 ± 34	629 ± 34	1001 ± 10
100	2046 ± 17	307 ± 64	302 ± 10	181 ± 5	686 ± 37	1225 ± 36
200	2133 ± 13	377 ± 48	309 ± 5	146 ± 11	625 ± 10	1226 ± 37

[a] Results of two independent experiments are shown (top and bottom). In all cases, numbers ± SEM shown come from duplicate RIA measurements for each treatment in triplicate, i.e., added to three tissue culture wells in the bioassay.

[b] Analysis of variance (EXBIOL) of all results showed a highly significant treatment effect; subsequent linear regression-analysis (BIOPROG) showed the results to be linearly distributed when relating effects and doses (see data in Fig. 1 for more evidence on this statement).

[c] Analysis of variance (EXBIOL) of all results showed no significant treatment effects.

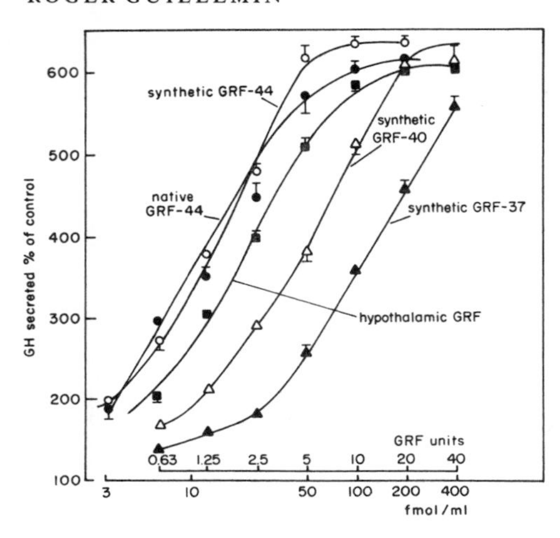

Fig. 8. Dose response curves for multiple doses of hypothalamic GRF, native hpGRF-44, synthetic GRF-40, synthetic GRF-37. The vertical bar on symbols represents standard error of the mean; when no such bar appears, standard error of the mean is no greater than the height of the symbol as drawn to indicate value of the mean response.

is taken as 100, in the assay described in Fig. 8 in which all the peptides were tested simultaneously on the same cell preparation, the potency of native or synthetic hGRF-40 is 30 (95% confidence limits: 25 and 37); the potency of native or synthetic hGRF-37 is 12 (95% confidence limits: 9 and 16). From calculation of the mean potency in 6 independent experiments hGRF-44 is 2.6 times more potent than GRF-40 with 95% confidence limits of 2.3 and 3.2.

These results led to calculating that one unit of GRF activity in the purified hypothalamic extract used as the reference standard corresponds to ca. 10 fmols GRF-44. Thus the extract of one rat hypothalamic fragment contains 350–500 femtomoles of GRF-44. Based on extensive studies (not shown here) of various extraction methods for fresh rat hypothalamic tissues, that figure was never seen to vary more than 2-fold.

IS THE ACTION OF GRF ON THE RELEASE OF GH DEPENDENT ON THE PRESENCE OF EXTRACELLULAR Ca^{2+}?

Data presented in Table 14 shows that a blocker of calcium uptake, $CoCl_2$ at 0.2 mM, inhibits partially and at 2.0 mM abolishes completely the response to hypothalamic or synthetic hGRF-40 or synthetic hGRF-44.

Table 14 *Effect of $CoCl_2$ on GH-releasing activity of hypothalamic GRF or synthetic GRF-40 and GRF-44*

Hypothalamic GRF GRF reference standard (in GRF units)	$CoCl_2$	GH released (ng/ml)	$CoCl_2$ (m*M*)	GH released (ng/ml)
0	0	900 ± 23	2	523 ± 6
0.63	0	1856 ± 45	2	513 ± 14
1.25	0	2520 ± 46	2	520 ± 6
2.5	0	3347 ± 44	2	543 ± 14
5.0	0	4673 ± 70	2	550 ± 23
10.0	0	5410 ± 65	2	613 ± 19
20.0	0	5580 ± 51	2	563 ± 22
40.0	0	5623 ± 67	2	560 ± 15
Synthetic hpGRF-40 (fmol/ml)				
0	0	1530	0.2	1050
6.3	0	2923 ± 17	0.2	1393 ± 46
12.5	0	4166 ± 14	0.2	1627 ± 17
25	0	5397 ± 16	0.2	2033 ± 21
50	0	6867 ± 46	0.2	2546 ± 13
100	0	8690 ± 12	0.2	3503 ± 79
200	0	8973 ± 51	0.2	3677 ± 59
400	0	9097 ± 14	0.2	3667 ± 67
Synthetic hpGRF-44 (fmol/ml)				
0	0	323 ± 7	2	150 ± 6
3.1	0	637 ± 20	2	150 ± 10
6.3	0	943 ± 27	2	150 ± 10
12.5	0	1273 ± 22	2	147 ± 18
25	0	1647 ± 37	2	157 ± 7
50	0	2017 ± 20	2	166 ± 9
100	0	2343 ± 20	2	167 ± 9
200	0	2383 ± 49	2	207 ± 7

IS THE ACTION OF GRF TO RELEASE GH MEDIATED BY THE ADENYLATE CYCLASE-cAMP SYSTEM?

Increasing cellular content of cAMP by adding 8Br.cAMP in doses ranging from 0.9×10^{-5} M to 2.4×10^{-2} M stimulates release of GH with the same slope of the dose-response curve as that obtained for increasing doses of hypothalamic GRF, native hGRF-44 or synthetic hGRF-40 (Fig. 9a). The same maximal effect (E_{max}) is reached (Fig. 6a) for all forms of GRF and for 8Br.cAMP. In presence of 8Br.cAMP (4×10^{-5} M to 4.10^{-3}) the GH-secretion stimulated by increasing

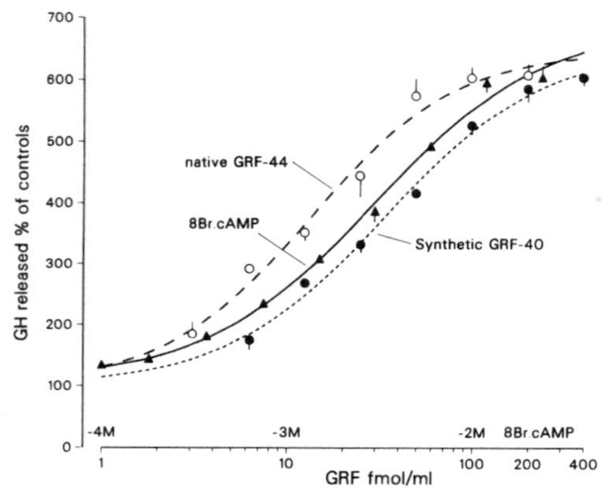

Fig. 9. a: Parallelism and identical E_{max} for the log dose-response curves for 8Br.cAMP, native hpGRF-44 and synthetic hpGRF-40.

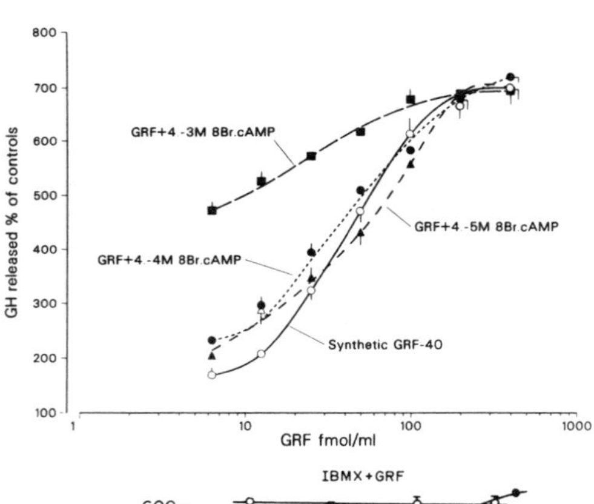

Fig. 9. b: Multiple doses of synthetic hpGRF-40 alone and in the presence of three different concentrations of 8Br.cAMP; additivity at the lower dose of both agonists but identical E_{max} for all agonists alone or in combination.

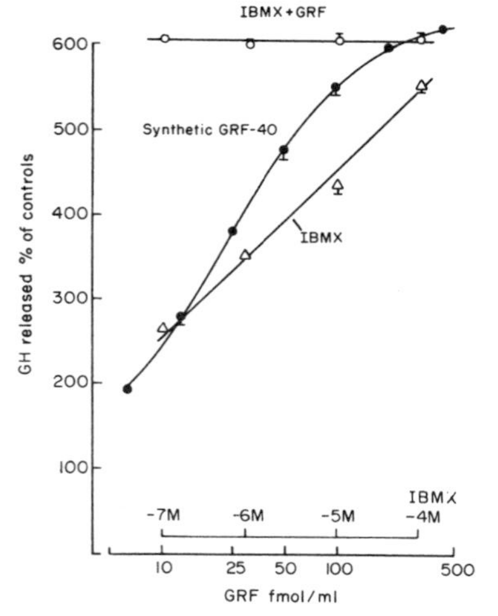

Fig. 9. c: Multiple doses of synthetic hpGRF-40, multiple doses of IBMX showing dose response curves and identical E_{max}; multiple dose of IBMX with a maximally stimulating dose of GRF (400 fmol/ml) show no increase in the E_{max} value as obtained for GRF alone.

Fig. 9. d: Same description and conclusion as in **c**, now for cholera toxin.

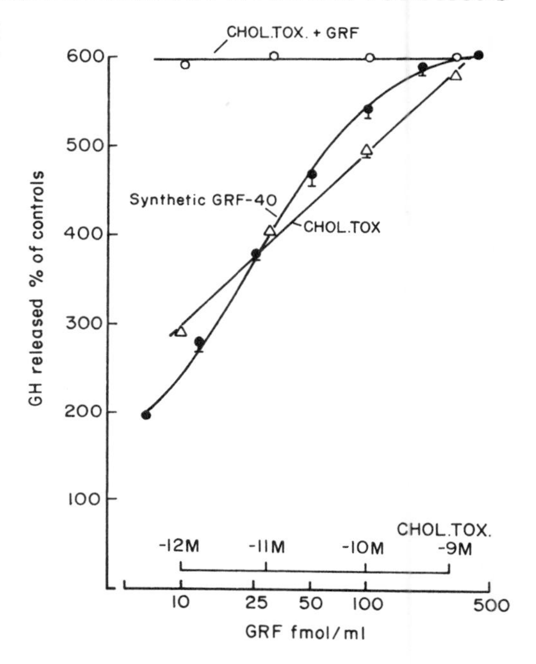

Fig. 9. e: Same description and conclusion as in **c**, now for forskolin. On all figures, the standard error of the mean value for any treatment is indicated by a vertical bar; when no such bar is shown, the standard error of the mean is smaller than the height occupied by the sign depicting that value of the mean.

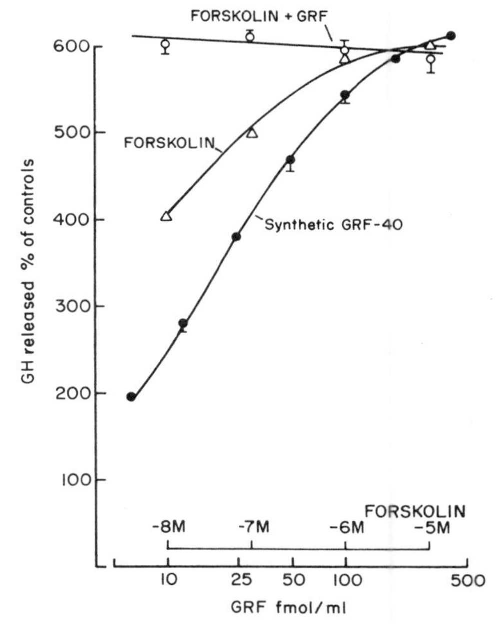

doses of synthetic hGRF-40 shows effect additivity of the two agonists, at the lower doses of GRF and at the higher doses of 8Br.cAMP, but in all cases the same E_{max} is reached in presence or absence of 8Br.cAMP (Fig. 9b).

The increasing availability of endogenous cAMP by adding the inhibitor of phosphodiesterases, IBMX (10^{-7} M to 10^{-4} M), stimulates secretion of GH as a function of the dose of IBMX added, with a slope of the dose-response curve statistically different from that obtained for hypothalamic GRF or synthetic hGRF-40 (Fig. 9c). When synthetic hGRF-40 is added at a maximally stimulating dose (400 fmols/ml) in presence of IBMX in increasing concentrations (1.10^{-7} M to 10^{-4} M), the value of E_{max} is never higher than that produced by the maximally stimulating dose of GRF alone (Fig. 9c).

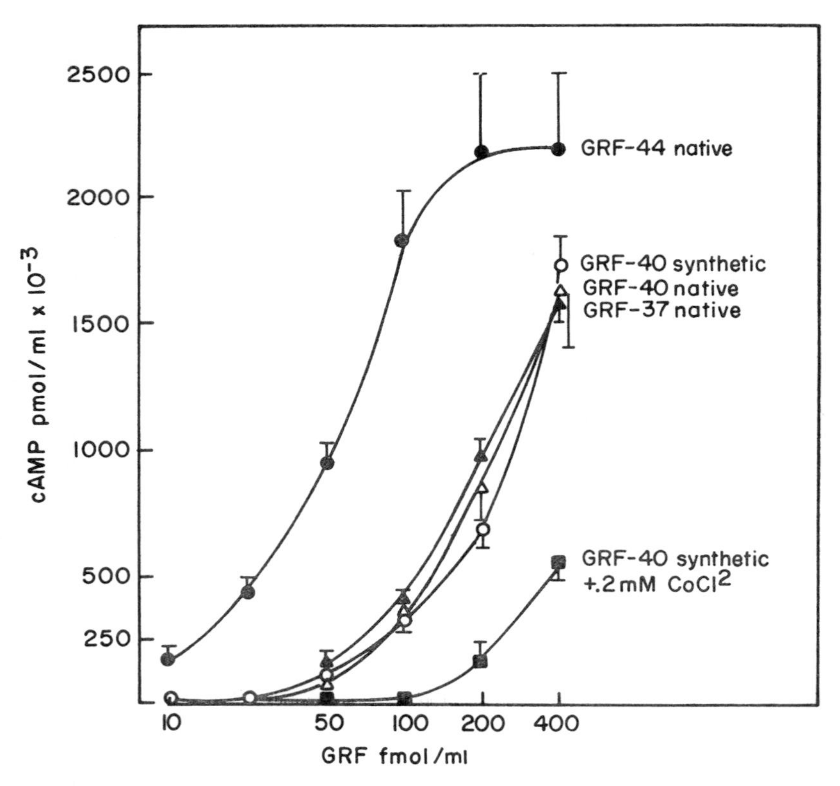

Fig. 10. cAMP released 4 hrs. in incubation fluids of monolayer pituitary cultures with multiple doses of several preparations of hpGRF, also in presence of 0.2 mM $CoCl_2$ added to GRF-40.

Figs. 11a, b. Results from two independent experiments showing the dose response curves to concentration of PGE_2 as shown, also responses to synthetic hpGRF-44 alone and in presence of multiple concentrations of PGE_2. Note the additivity of the maximal effect due to each agonist. Standard error of the mean shown by a vertical bar when that standard error is greater than the height of the sign showing the value of the mean.

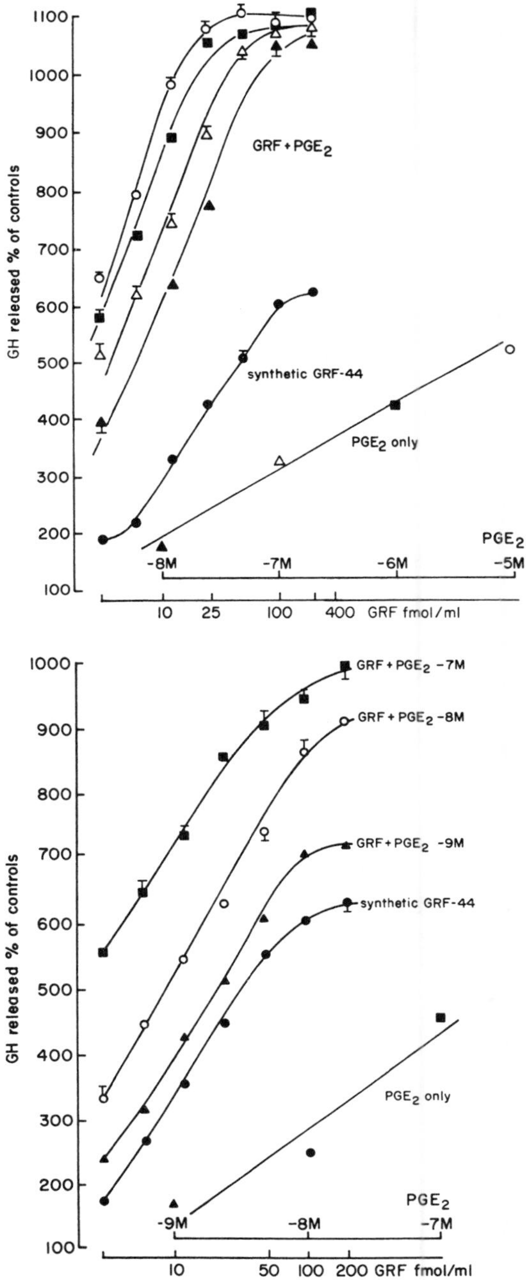

Similarly, cholera toxin (10^{-9} M to 10^{-12} M), an activator of the regulatory subunit, and forskolin (10^{-5} M to 10^{-8} M), a stimulator of the catalytic subunit of adenylate cyclase, stimulate secretion of growth hormone with dose-response curves statistically different from that of synthetic hGRF-40 (Figs. 9d,e); when the same multiple doses of cholera toxin or forskolin are added, in presence of a maximally effective dose of synthetic hGRF-40, the maximal response (GH secretion) to all treatments is not different from the E_{max} observed with GRF alone (Figs. 9d,e).

Direct measurement of cAMP released in the culture fluid by the pituitary cells shows an increase as a function of the dose of GRF added to the monolayer pituitary culture. Efflux of cAMP reaches a plateau for the same doses of hGRF-44 that also yield E_{max} for GH secretion. The increases in released cAMP parallel the increases in released GH as a function of the potency of GRF added (Fig. 10) or available (see for instance cAMP released by the same doses of hGRF-40 in absence or presence of $CoCl_2$ (Fig. 10) comparing with the effects, in the same experiment, of $CoCl_2$ on the release of GH, Table 13, panel b).

GRF AND PROSTAGLANDIN PGE_2 STIMULATE SECRETION OF GH BY DIFFERENT PATHWAYS AND MECHANISMS

In contradistinction to the results obtained with 8Br.cAMP, IBMX, cholera toxin and forskolin, studies of interrelationships between prostaglandin PGE_2 and GRF give entirely different results. The dose response curve to PGE_2 is totally divergent from that observed for hypothalamic GRF or synthetic hGRF-44 (Figs. 11a,b); moreover, PGE_2 E_{max} never reaches that due to GRF (Figs. 11a,b) even at the highest tolerable doses of PGE_2 (10^{-2} M). When multiple doses of synthetic hGRF-44 are studied on GH secretion in presence of PGE_2 (10^{-8} M to 10^{-5} M) a remarkable additivity of effects is observed at all doses, with values for E_{max} of the combined treatments far greater than those regularly observed for GRF alone (Figs. 11a,b).

HOW RAPID IS THE EFFECT OF GRF IN ELICITING RELEASE OF GH AND IS IT DEPENDENT ON THE SYNTHESIS OF SOME INTERMEDIATE PROTEIN?

Results presented in Fig. 12 from one perifusion experiment with dispersed pituitary cells show that the effect of hypothalamic GRF or of synthetic hGRF-44 to stimulate release of GH is demonstrable

Fig. 12. Rapidity of the pituitary response to hypothalamic GRF or synthetic GRF-44 is shown in a perifusion system using dispersed pituitary cells. Each fraction collected is 250 μl, for 33.8 seconds; total duration of each GRF pulse is 155 seconds.

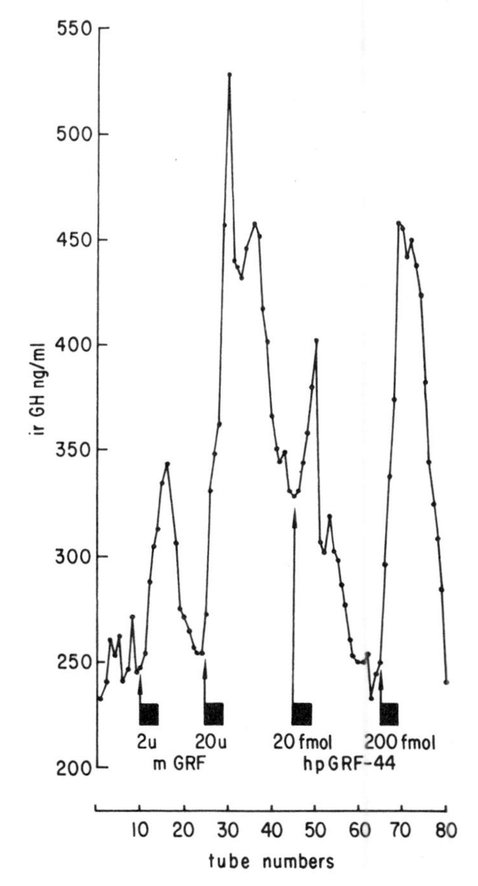

in ca. 30 seconds following the contact of GRF with pituitary cells. In this perifusion system the effect of GRF is relatively short-lived, the duration of effect being related to the dose of GRF for identical pulse durations. The effect of synthetic hGRF-40 on the release of GH is not modified by doses of cycloheximide as high as 100 μg/ml, added 2 hours prior to GRF (data not shown here). These doses are well above those necessary to inhibit protein synthesis in the same *in vitro* system (Vale *et al.*, 1968).

ANTAGONISM BETWEEN GRF AND SOMATOSTATIN

As shown in Figs. 13a and 13b, somatostatin-28 or somatostatin-14 inhibit the response to hypothalamic GRF or native hGRF-44 in a

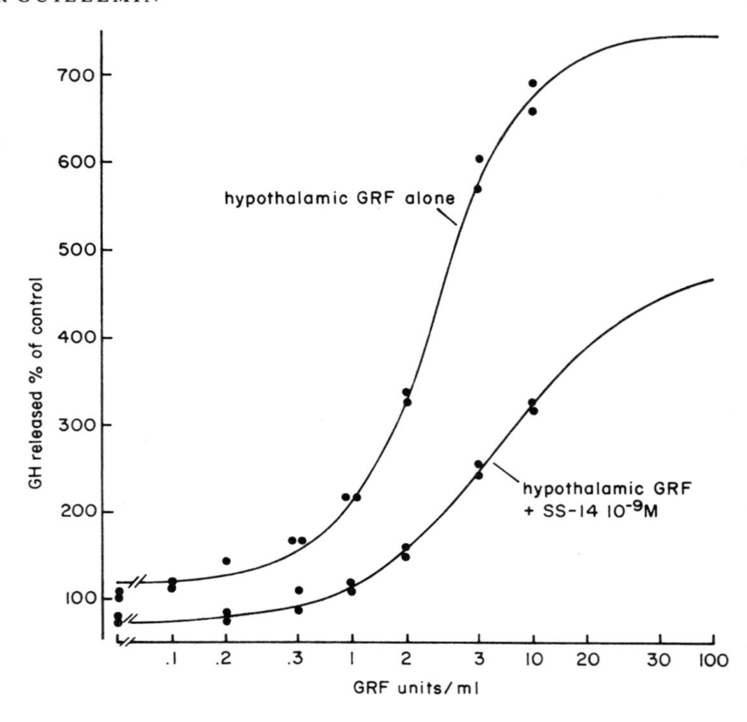

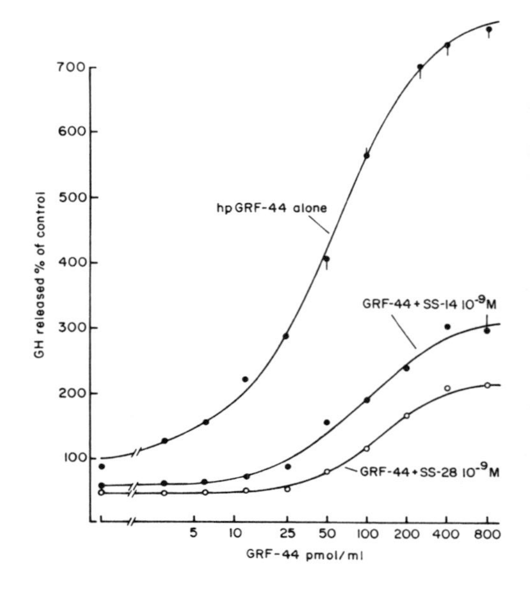

Figs. 13a, b. Somatostatin-14 or somatostatin-28 inhibit the response to hypothalamic GRF or native hpGRF-44 in typical noncompetitive antagonism. Results of two independent experiments. Symbols show actual experimental data; lines are the theoretical curves computer-calculated and drawn, from the 4-parameter logistic equations for each set of data; curves shown here are drawn without constraints (program ALLFIT).

typical non-competitive relationship. Analysis of the dose-response curves by the 4-polynomial equation of De Lean (1978) using the program ALLFIT shows that values for parameter *b* (slope) are statistically identical (for each set of curves), and so are values for parameter *c* (ED_{50}) for each set of curves. Similarity of the ED_{50} of the agonist (GRF) in presence or absence of the antagonist (somatostatin) is one of the main criteria for non-competitive inhibition. Values for parameter *a* (response at dose 0 of the agonist) and *d* (E_{max}) are different when comparing each set of curves: heterogeneity of the values for *a* indicates that somatostatin also affects the basal secretion of GH by the pituitary cells *in vitro*. Heterogeneity of the values for *d* is another criterion of non-competitive antagonism; the antagonist (somatostatin) acts at some locus other than the receptor of GRF to prevent the full activity of the agonist. The greater inhibition by somatostatin-28 than that due to (equimolar amounts) of somatostatin-14 (Fig. 13*b*) reflects the greater potency of somatostatin-28 when compared to the tetradecapeptide as originally reported by us in *in vitro* and *in vivo* systems (Brazeau *et al.*, 1981*a*). These data show that the biological activity of hypothalamic GRF is qualitatively indistinguishable from that of any of the three characterized forms of tumor-derived GRF or of their synthetic replicates. This conclusion has now been reinforced and extended with availability of synthetic replicates of hypothalamic porcine, bovine and murine GRFs. The slopes of the dose-response curves are identical and so are the values for maximal stimulations (E_{max}) obtained with the hypothalamic or the tumor-derived preparations of GRF, native or synthetic. What differ are the specific activities (number of biological units/mole) of these various preparations. All evidence points to hGRF-44 as being the primary form of human GRF: it is statistically more potent than any other form; it exists as a C-terminal amide, the form in which many neuropeptides (TRF, LRF, CRF, bombesin, substance P, vasopressin, oxytocin, etc.) and peptide hormones (α-MSH, secretin, cholecystokinin, gastrin, gastrin releasing peptide, etc.) have been characterized and have maximal activity or are exclusively active (TRF, LRF, CRF, vasopressin, etc.). Moreover, the amino acid sequence of GRF-44 shows that GRF-37 and GRF-40 could be generated from GRF-44, by cleavage at the NH_2-terminal side of the Arg residues in position 38 and 41, as has been found to be the case for dynorphin (1–8) (Minamino *et al.*, 1980) from dynorphin and for dynorphin B (rimorphin) from the COOH-terminal of the β-neo-endorphin precursor (Kadikani *et al.*, 1982). Finally, and as shown above (see page 11), knowledge of the primary structure of the precursor of hGRF, pre-pro-GRF, deduced from its encoding cDNA, confirms the complete sequence of GRF-44 with evidence of an amidating signal. As mentioned above, the same conclusion has been published by Mayo *et al.* (1983) showing by molecular cloning the same sequence of the precursor of GRF using, as source of GRF-mRNA, tissue from the pancreatic tumor from Thorner's patient in which only GRF-40 was found and characterized (Esch *et al.*, 1982; Rivier *et al.*, 1982).

The high specific activity (number of biological units per mole) of hGRF-44 is worthy of comment. The data reported here show hGRF-44 to be significantly active in releasing GH in the monolayer culture assay at $\leqslant 3$ fmol/ml or 3.10^{-12} M. In the dispersed pituitary cell perifusion assay the minimal active dose is $\leqslant 5$ fmol/250 μl/30 seconds. The potency of hGRF-44 is thus not only in the same range as that of TRF, LRF, CRF, but even greater, in comparable assay systems *in vitro*. Such high potency of the material, along with its specificity for influencing the secretion of only GH, speaks in favor of its physiological significance as a GH-releasing factor.

The early results reported here show that extracellular Ca^{2+} is necessary in the mechanism of action of GRF. The evidence presented here is best explained by proposing that the adenylate cyclase-cAMP system is involved in the mechanism of action of GRF in stimulating the secretion of GH. The biological system used here is not a homogenous population of somatotrophs; therefore, the only specificity attributable to the results obtained stems exclusively from the specificity of GRF in acting only on somatotrophs (since there is no evidence that it stimulates secretion by the pituitary of anything else other than GH). Cronin *et al.* (1982) have recently reported comparable preliminary results.

Recently Lewin *et al.* (1983) reported that synthetic hGRF-44 increases in a dose-dependent manner and with a half-maximal effect at 35 ± 8 pM the activity of a cAMP-dependent protein kinase present in purified hog anterior pituitary granules. Analysis of the phosphorylation kinetics suggested that the peptide did not significantly change the reaction's V_{max} but produced a major increase in the enzyme affinity for cAMP: the apparent K_M for the nucleotide decreasing from 700 nM in control, unstimulated condition, to 15 pM in the presence of 100 pM GRF-44 (see Fig. 14). Less potent analogs of GRF-44 such as GRF-37 or [Phe^1]-GRF-40 had lower effects on the cAMP-dependent protein kinase of the pituitary granules in relation to their lower potency in the bioassay for GH release; an analog, GRF(2-40), inac-

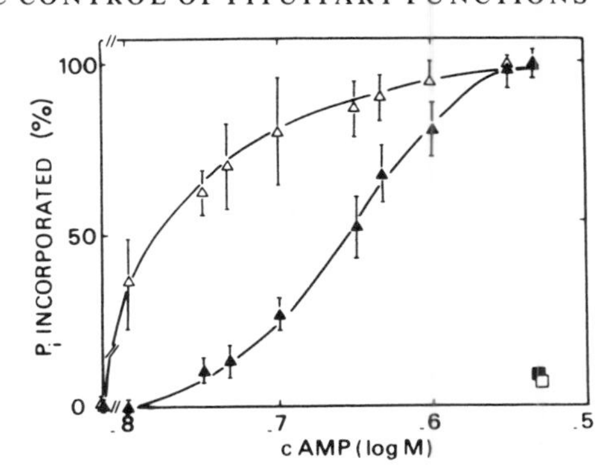

Fig. 14. Histone phosphorylation by purified hog pituitary granules at different concentrations of cAMP without (closed triangle) and with (open triangle) 100 pM GRF. Mean ± SEM from 4 different preparations. P_i values shown by the squares are in the presence of the Walsh reagent at the concentration of cAMP shown on the x-axis. (From Lewis et al., 1983).

tive in stimulating the secretion of GH, had no evident effect on the phosphorylating enzyme of the pituitary granules at doses as high as 0.1 μM. The authors conclude by suggesting that GRF stimulates secretion (release) of growth hormone by activating exocytosis through a phosphorylation mechanism mediated by a granular receptor coupled with a cAMP-dependent protein kinase.

Using techniques of immunocytochemistry on tissue sections prepared for electron microscopy, and with several antisera raised against hGRF provided by our laboratory, Morel *et al.* (1984) concluded that they could visualize immunoreactive GRF grains in normal pituitary somatotrophs of Man and Rhesus Monkey. The immunoreactive GRF grains are located on GH-secretory granules as well as in the cell nucleus. No image is seen in rat pituitary preparations, in agreement with the fact that the hGRF antisera prepared by us do not read rat GRF. However, when hGRF is injected intravenously into rats (and we know that hGRF is highly active in stimulating the secretion of growth hormone in the rat), Morel *et al.* report that they can see immunoreactive GRF grains, again located on GH-secretory granules, as early as 3 min. after injection of hGRF, their number increasing at first and then decreasing as the somatotrophs become degranulated as they secrete GH, while the immunoreactive GRF grains are seen in the nucleus from 3 min. up to 30 or 60 min. after injection. No such images are seen with hGRF(1-28), a biologically active fragment of GRF not recognized by the antiserum used above. Similarly, no such images are seen following injection of hGRF(28-40), a biologically inactive fragment recognized by the antiserum provided by us and used here.

These elegant conclusions are surprisingly well in agreement with the biochemical results of Lewin *et al.* (see above) on the modus and locus of action of GRF at the level of the GH-secretory granules; they also are in accord with results of the effect of GRF on GH-mRNA levels (see below). In view of the complexity and unusual methodology involved in these studies, the conclusions of Morel *et al.* would gain weight and acceptance if they could be confirmed by others and/or by the use of other methods such as the colloidal gold labelling of antibodies.

The rapidity of action of GRF either of hypothalamic origin or as hGRF-44 showing activation of the somatotrophs in seconds is in keeping with similar characteristics for the other hypothalamic releasing factors, TRF, LRF, CRF. Frohman *et al.* (1980), using perifusion of whole rat pituitaries, had reported that the GRF material they had purifed from a human carcinoid that had caused acromegaly also elicited rapid release of GH. The frequency of sample collection, however, was far slower than the one used here (5 min. *vs.* 30 sec.) for a 15-min. GRF-pulse, *vs.* 120-sec. GRF pulse in our own experiments. In that same report Frohman *et al.* (1980) elaborated on various characteristics of the material with GRF-activity they had purifed from several human tumors which had caused acromegaly. Many of their proposals regarding molecular size, significance of the NH_2- and COOH-terminals, discrepancies with activity of hypothalamic extract, conclusions as to existence of precursor forms, etc., of their active material are at variance with our present knowledge of the fully characterized hGRF. It is always difficult to draw such conclusions when dealing with non-homogenous materials. The merit of Frohman *et al.* in that report (1980), as well as that of other reports (UzZafar, 1979; Beck *et al.*, 1973; Leveston *et al.*, 1981; Caplan *et al.*, 1978), is in having brought forth early evidence for GRF-activity in extracts of peripheral tumors accompanying acromegaly, thus leading to the then novel concept of ectopic production of GRF-like substances.

The non-competitive nature of the inhibition of the activity of hypothalamic GRF or hGRF-44 by somatostatin, though never reported as such by others (Cronin *et al.*, 1982; Frohman *et al.*, 1980), was not an unexpected finding. There is no evidence that somatostatin or any of its many analogs ever behaved as a partial agonist (on the release of growth hormone); indeed, the latest proposal (Reyl and Lewin, 1981) for the sub-cellular mechanism of action of somatostatin would lead one to expect results and conclusions as to a non-competitive mode for the antagonism between somatostatin and GRF.

With the availability of highly purified preparations of somatomedin C and IGF-I and IGF-II, we have recently demonstrated that these peripheral GH-induced polypeptides can inhibit directly at the pituitary level the GH-releasing activity of GRF (Brazeau *et al.*, 1982*b*). The inhibitory activity is seen at concentrations of 0.5–10 ng/ml (1×10^{-10} M, 1.3×10^{-9}) in either short-term (3–4 hr.) or long-term (24 hr.) incubation. These results are in agreement with Berelowitz *et al.* (1981), who proposed a direct negative pituitary feedback effect of somatomedin-C, but at variance with their data showing, under their experimental conditions, that a long-term ($\geqslant$ 24 hr.) contact with somatomedin-C would be necessary. In our experiments, IGF-II was also active as an inhibitor of GRF, though less potent than IGF-I. The effect of the somatomedins is specific for the secretion of growth hormone and is not duplcated by other growth factors such as epidermal growth factor (EGF) or fibroblast growth factor (FGF).

The effect of hGRF-44 was studied concomitantly on GH-release and GH-mRNA levels in normal pituitary cells in monolayer culture as above and in GH_3 cells, a cell-line known to secrete GH and PRL spontaneously, and extensively used by others to study the mechanism of the secretion of GH or of PRL with various secretagogues (TRF) or inhibitors (somatostatin). The cytoplasmic dot hybridization technique (White and Bancroft, 1983) was used to examine the effect of GRF. Pituitary cells incubated for 24 hr. with 25–44 fmol GRF had significant increase in GH-mRNA levels. Maximal GH-mRNA levels (2.5-fold increases over controls) were noted following a 72-hr. incubation of normal pituitary cells with 10^{-9} M GRF. In a similar experimental paradigm GRF did not stimulate PRL release or relative PRL-mRNA levels. GRF does not elevate GH-mRNA or PRL-mRNA levels in GH_3 cells; GRF does not stimulate either the secretion of GH by GH_3 cells (Gick *et al.*, 1983). These observations and conclusions have recently been confirmed by Barinaga *et al.* (1983) studying the transcriptional regulation of the GH gene by GRF.

In closing this section on *in vitro* studies with GRF, it is intriguing to mention the first, and so far only, evidence of a secretory activity of GRF not at pituitary level. Zeytin and Brazeau (1984) have shown that high doses of hGRF-44 and rGRF increase in a dose-dependent manner the secretion of neurotensin (ED_{50} 10^{-9} M), calcitonin (ED_{50} 10^{-10} M) as well as cAMP content and release by a cell-line originating from a medullary thyroid carcinoma (MTC). Because of the homogeneity of such a cell-line, and because the various parameters studied so far on the GRF-stimulated secretion of neurotensin parallel remarkably similar effects on the GRF-induced secretion of GH by pituitary cells, the MTC cell-line may be an elegant model to use or consider.

In vivo studies with synthetic replicates of GRF in laboratory animals

With the availability of the synthetic replicates of GRF, extensive *in vivo* studies could be initiated. Initially it was necessary to establish the dose-response relationship between GRF and plasma GH levels, as well as the specificity of GRF. Fig. 15 illustrates the dose-response relationship between hGRF-44 and plasma GH levels in anesthetized male rats. More detailed investigations of the time course as well as the minimum and maximum effective dose of GRF necessary to stimulate GH secretion (Wehrenberg *et al.*, 1982*a*; Wehrenberg *et al.*, 1983) have been reported. These results have shown that in rats the pituitary GH response to GRF is very rapid and of short duration, that is, the pituitary GH response is maximal at 3–5 minutes and plasma GH concentrations are returning to base-line within 15 minutes. The minimum effective dose is approximately 100 ng/kg and the lowest dose producing a maximal effect approximately 10 μg/kg.

Synthetic hGRF-44 is a potent secretatogue of GH in the dog (Guillemin *et al.*, 1982). In complete agreement with *in vitro* observations (see Table 13), GRF administration produced no changes in the plasma concentrations of LH, FSH, TSH, PRL and corticosterone (Wehrenberg *et al.*, 1982*a*).

For obvious reasons, animals anesthetized with sodium pentobarbital are not the best model for studies designed to investigate mechanisms regulating GH secretion. Therefore studies were initiated in conscious, freely-moving rats outfitted with chronic indwelling venous catheters and maintained in isolation chambers. In animals so prepared the administration of GRF during the interval between spontaneous GH pulses (Tannenbaum and Martin, 1976) failed to elicit a consistent increase in plasma GH concentrations (Wehrenberg *et al.*, 1982*c*). Of the 15 rats so treated, only 5 demonstrated an increase in GH comparable to what we had observed in anesthetized rats. Fig. 16 illustrates 4 individual examples of the inconsistency observed in the response. In contrast, GRF consistently elicited a dramatic and immediate increase in plasma GH when these animals were pretreated with antibodies raised against somatostatin, the hypothalamic inhibitor of GH release (Fig. 17).

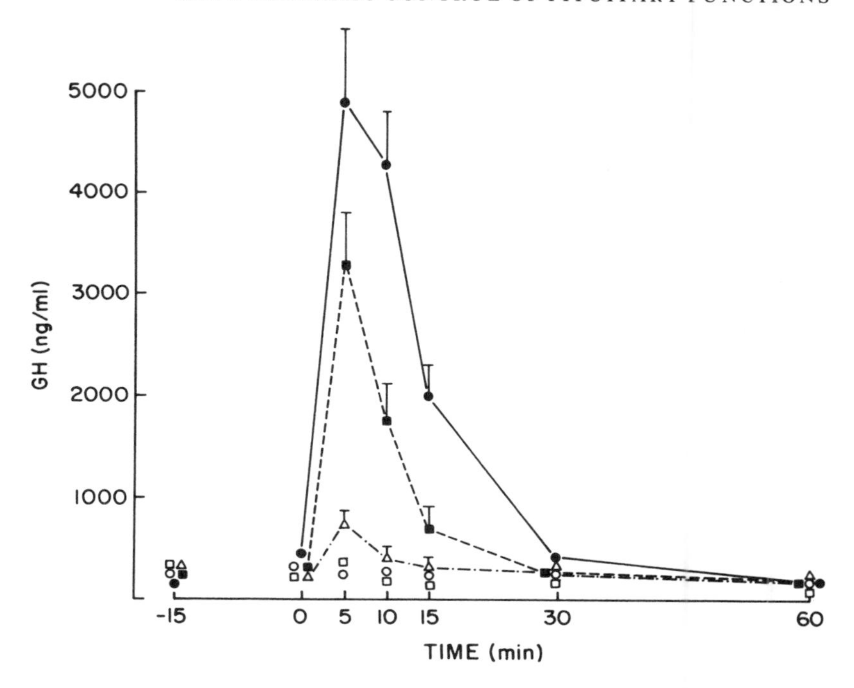

Fig. 15. Plasma growth hormone levels in response to the intravenous administration of saline (□) and of synthetic hpGRF-44 at 0.01 μg (○), 0.1 μg (△), 1.0 μg (■) and 10.0 μg (●). Male Sprague-Dawley rats weighting 280 to 320 g were anesthetized with sodium pentobarbital (50 mg/kg) intraperitoneally) at time −30 minutes. Samples (0.2 ml) were drawn by venipuncture at times indicated; saline or hpGRF-44 was administered immediately after time 0. Four animals were used for each treatment. Results are shown as the mean of responses, with the vertical line representing the standard error of the mean.

To establish further the dynamic role of somatostatin in modulating the GH response to GRF, rats were subjected to a 72 hr. fast, a procedure which increases endogenous somatostatin (Tannenbaum *et al.*, 1978). In such animals GRF *i.v.* administration fails to increase plasma GH concentrations; however, a response could regularly be elicited with GRF in these fasting rats if they were first pretreated with somatostatin antibodies (Wehrenberg *et al.*, 1982*b*).

As evidenced by Figs. 16 and 17, the concentration of plasma GH during spontaneous pulses can approach the concentrations obtained following exogenous administration of GRF. A possible criticism of studies involving the exogenous administration of GRF is that one cannot be certain whether the increase in plasma GH would be caused

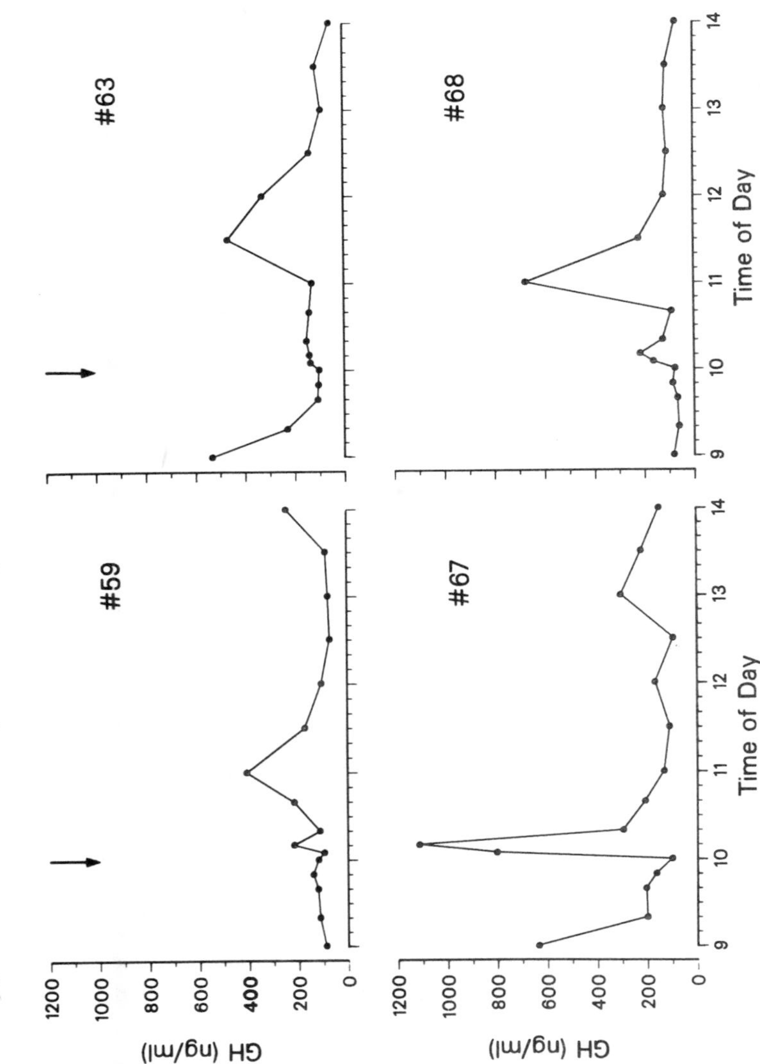

Fig. 16. The effect of 10 μg hpGRF-44 or -40 (i.v.) on GH secretion in 4 individual, conscious, freely-moving male rats. Injections (indicated by arrows) were made at a time known to be between spontaneous GH pulses. Note the absence of response in rat #63 and the partial response in rats #59 and #68 as compared to the response in rat #67.

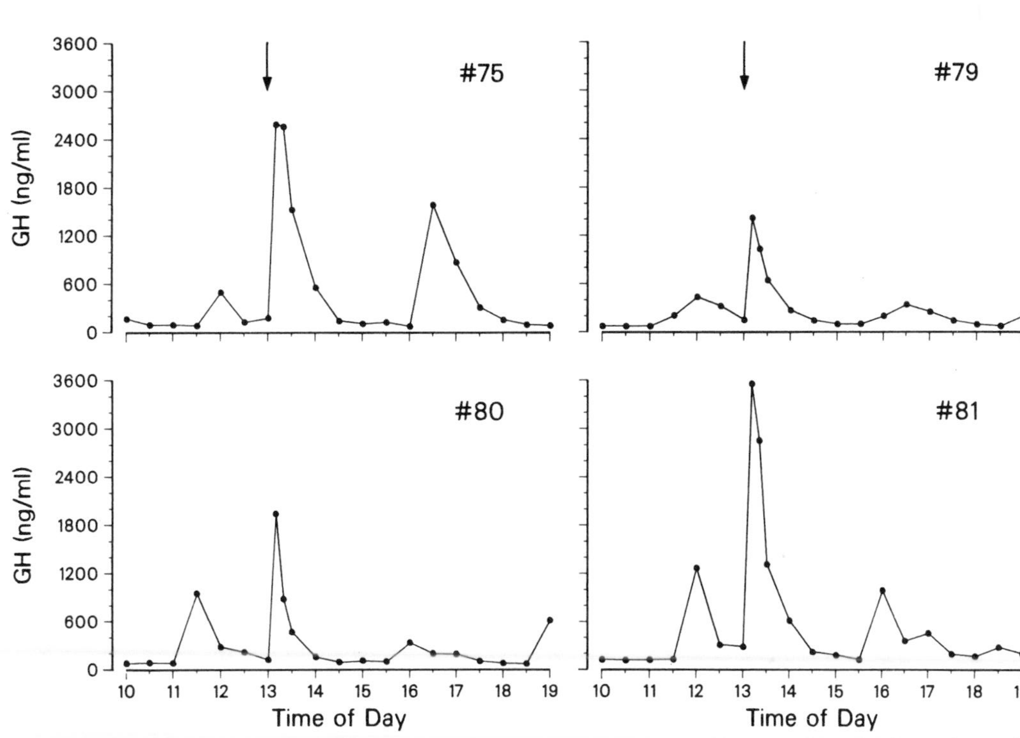

Fig. 17. The effect of 1 μg hpGRF-44 or -40 on GH secretion in 4 individual, conscious, freely-moving male rats pretreated with 5.8 mg protein of antibodies against somatostatin. Injections (indicated by arrows) were made at a time known to be between spontaneous GH pulses. Note the change in dose of hpGRF administered and scale of GH concentrations as compared to Fig. 13.

by GRF injection or a spontaneous GH pulse. During the course of the isolation and purification of GRF, Richard Luben developed several monoclonal antibodies against rat hypothalamic GRF using *in vitro* techniques (Luben *et al.*, 1982). Passive immunization of rats with these antibodies has been shown specifically to inhibit the pulsatile secretion of GH presumably by neutralizing hypothalamic GRF (Wehrenberg *et al.*, 1982*c*). One of these monoclonal antibodies against rat hypothalamic GRF does not recognize the GRF isolated from the human sources and thus permits an answer to the criticism of spontaneous versus stimulated increases in plasma GH. Indeed, combining the passive immunization treatment of the monoclonal antibodies with treatment with somatostatin antiserum results in a unique animal model. This model is ideal in that it represents an immediate, non-invasive, yet reversible, functional lesion of the hypothalamo-pituitary axis which is specific for only endogenous somatostatin and rat GRF. Using this model, the pituitary GH response to human GRF follows a dose-response relationship (Wehrenberg *et al.*, 1984*b*); this response is virtually unchanged over time to either a low (0.25 μg) or a high (5 μg) dose when administered at hourly intervals (Fig. 18).

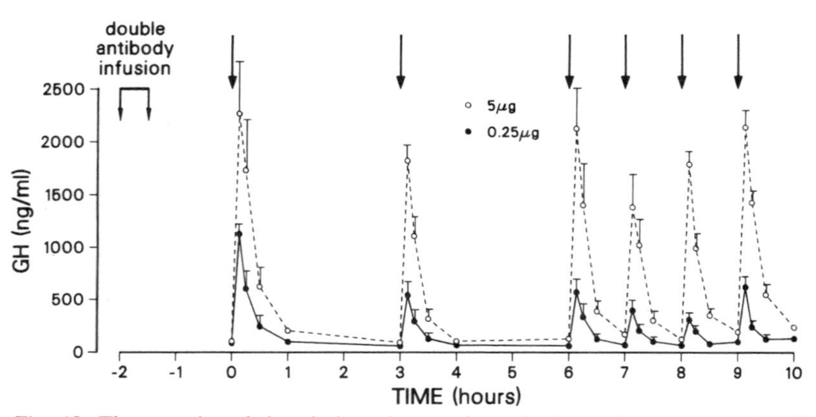

Fig. 18. The capacity of the pituitary in conscious, freely-moving male rat (n = 6) to secrete GH in response to repeated intravenous injections of a moderate (0.25 μg; ●) and maximal (5 μg; ○) dose of hpGRF-40. Two hours before the first injection rats were treated with antiserum against somatostatin and monoclonal antibody against rat hypothalamic growth hormone-releasing factor. Arrows indicate the injection of synthetic hpGRF-40. Data points represents the mean GH concentration; the vertical bars represent the SEM.

While the fragments GRF-37, GRF-40 are less potent than GRF (i.e. GRF-44-NH_2) when tested in the *in vitro* system (see above), all three peptides with GRF activity are statistically equipotent when tested *in vivo* (Wehrenberg *et al.*, 1983).

GRF is active in stimulating GH secretion in rats with functional or anatomical lesions of the central nervous system that have been shown to inhibit endogenous growth hormone secretion (Wehrenberg *et al.*, 1984*a*). These animal models include electrolytic lesions of the ventromedial hypothalamus, a chemical lesion of the arcuate nucleus induced by neonatal treatment with monosodium glutamate, a functional lesion of catecholamine synthesis by pretreatment with α-methyl-p-tyrosine and a functional lesion of catecholamine storage with reserpine. These results further demonstrate that GRF is indeed a secretagogue of GH by acting directly on the pituitary and not at some proximal locus within the central nervous system.

In 1946 Hans Selye stated that 'while the pituitary is actively engaged in increased corticotropin hormone production it is apparently less capable of elaborating growth hormone, prolactin and gonadotropic hormones.' This 'shift in hormone production', as described by Selye, is now well recognized. Two examples include the stress-induced changes in gonadal and thyroid function (Ducommun *et al.*, 1967; Sachar, 1975). One possible mechanism that might account for this 'pituitary shift' would be an interaction of the pertinent releasing factors at the pituitary level. With the synthetic replicates of luteinizing hormone-releasing factor (LRF), corticotropin-releasing factor (CRF), thyrotropin-releasing factor (TRF) and growth hormone-releasing factor (GRF) now available, we decided to examine that hypothesis.

This was tested *in vitro* and *in vivo* using a 2^4 factorial experimental design (Table 15). This experimental design allows for the evaluation of both the main treatment effects of the hypothalamic releasing factors as well as all of the possible interactions between them. Significant main treatment effects both *in vitro* and *in vivo* were: LRF on LH and FSH, CRF on ACTH and β-endorphin, TRF on TSH and GRF on GH. These results confirm the specificity of the four releasing factors on their respective target cells. There were no significant interactions between any of the releasing factors on anterior pituitary hormone secretions (Wehrenberg *et al.*, 1984*a*). These results suggest that the changes in pituitary secretion that are observed in health and disease are not due to interactions between the hypothalamic releasing factors at the level of the pituitary, but rather to the second-

Table 15 *Experimental design of a 2^4 factorial experiment to study interactions between releasing factors*

LRF(L)	CRF(C)	TRF(T)	GRF(G)
0*	0	0	0
0	0	0	1
0	0	1	0
0	0	1	1
0	1	0	0
0	1	0	1
0	1	1	0
0	1	1	1
1	0	0	0
1	0	0	1
1	0	1	0
1	0	1	1
1	1	0	0
1	1	0	1
1	1	1	0
1	1	1	1

*0 = no; 1 = yes

ary interactions that modify pituitary activation or response. These results also indicate that the clinical pituitary reserve test can be expanded to include, as a single bolus mixture, all four hypothalamic releasing factors, since any lack of response will reflect a pituitary problem and not an interaction of the secretagogues administered.

IN VIVO EFFECTS OF GRF OTHER THAN HYPOPHYSIOTROPIC

In acute toxicity studies conducted to satisfy requirements of the FDA for ultimate clinical applications of GRF, we have found GRF to be an unusually innocuous substance. Intravenous bolus injection of up to 2 mg synthetic GRF in monkeys (7.5–9.5 kg b.w.) or up to 1 mg in rats produces no obvious behavioral effect, save perhaps for a mild tranquilization of the animals. At these extremely high doses there is in rats a transient (5 min.) minimal fall of blood pressure (5–10 mm Hg), as shown by direct recording of aortic pressure in animals with chronic catheters (A. Briskin). This is accompanied by a rise in pulse rate, probably reflexively induced. Acute *i.v.* injection of these large doses of GRF produces no change in blood-sugar levels. Intra-cerebral injection of GRF (1–10 μg) in the lateral or third ventricle again produces little obvious effect, except again a general quietening of the animals.

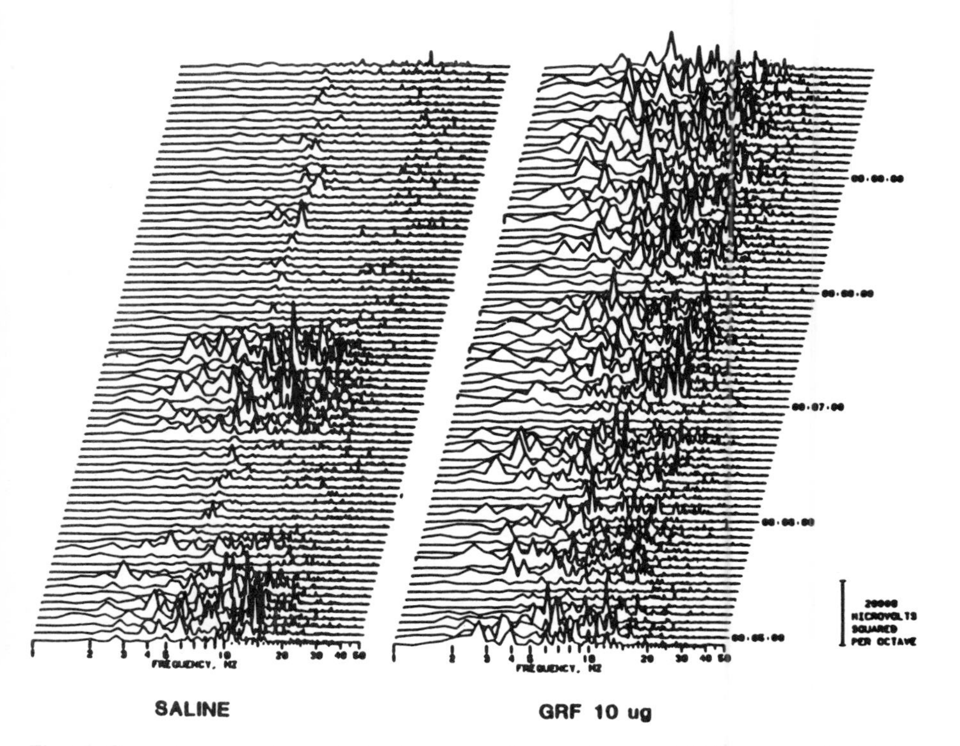

Fig. 19. Computer-generated analysis of compressed spectral array of EEG recorded for 15 min. after i.c.v. injection of saline or GRF-44. (Ehlers *et al.*, unpublished results.)

In early studies Ehlers *et al.* (to be published) observed with continuous EEG monitoring that intra-ventricular injection of hGRF-44 tends to maintain the EEG in the stage of slow-wave sleep (Fig. 19). Other computer treatment of the EEG (power spectral band time series) shows GRF to increase EEG stability.

This overall tranquilizing effect of large doses of GRF, the complete opposite of the excitatory effects of CRF, a peptide of similar size, is reminiscent of the clinical comportmental makeup of all acromegalics, described by physicians as quiet, unhurried, 'the smiling apathy' of the acromegalics, in the terms of Bleuler (1951). Mandell (1983) has gone so far as to propose that GRF could be the substrate of the neuroendocrine trophotropic functions of the hypothalamus, while CRF would be the ergotropic counterpart, in the terminology and concept of Walter Hess (1948).

Localization of GRF-containing cells using immunohistochemistry

The first and basic question addressed was whether there would be any immunological community between hpGRF and hypothalamic GRF. Indeed we found that antibodies raised against hpGRF-40-OH and hpGRF-44-NH_2 specifically stained discrete neurons in primate hypothalamus, the topography and organization of which were characteristic of a neuronal system producing a hypophysiotropic releasing factor. Structures stained with hGRF antibodies were, both in human and monkey, dense bundles of fibers in the median eminence that terminate in contact with portal capillaries and immunoreactive cell bodies that were found essentially in the mediobasal hypothalamus, mainly in the infundibular arcuate nucleus. The ventromedial nucleus contained few neurons, or none at all, according to the samples. These cells bodies were found in small amounts and inconsistently in the monkey, while they were invariably numerous and intensely immunoreactive in humans (Figs. 20, 21) (Bloch *et al.*, 1983*a,b*).

Controls of inhibition by hGRF or fragments thereof demonstrated the specificity of the immunostaining in both species and also supplied information about antigens recognized by these antibodies. Antibodies against hGRF1-40 were inhibited by hGRF1-40, hGRF1-44-NH_2 and hGRF28-44-NH_2, indicating that this antibody recognizes both hGRF1-40 and hGRF1-44-NH_2 in their C-terminal region. Antibody against hGRF1-44-NH_2 was inhibited by hGRF1-44-NH_2 but not by hGRF1-40 or hGRF1-44-OH, showing that it was specific of the 1-44 amidated full-length form of GRF in its C-terminal ending. Since these antibodies recognized hGRF in two different epitopes located outside sequences common with other neuropeptides (PHI, VIP, etc.—see Table 9), it appears that the staining in the brain was indeed due to genuine GRF or related metabolic compounds. Stainings of neurons with antibodies specific for hGRF1-44-NH_2 showed the presence of the full-length form of amidated GRF in the brain, in accordance with radioimmunological detection in human hypothalamus of a peptide indistinguishable from hGRF1-44-NH_2 and its subsequent chemical characterization as hGRF1-44-NH_2.

Neurons containing hGRF immunoreactivity give projections in the anterior hypothalamus, especially in the paraventricular nucleus. Some fibers terminate as perisomatic endings, suggesting that, in addition to its neuro-humoral role, GRF could be involved in interneuronal relationships. Comparative topographical studies showed that

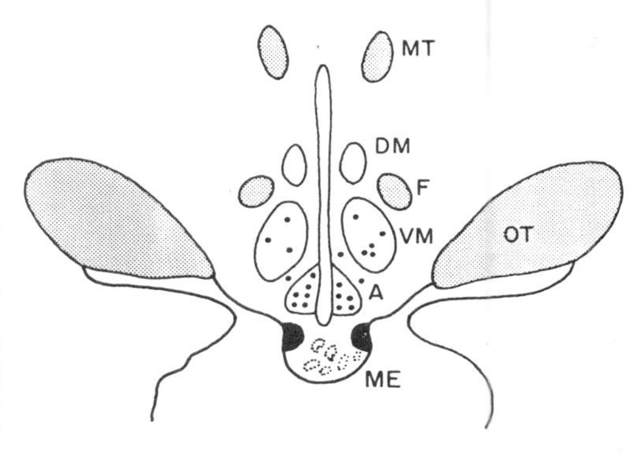

Fig. 20. Topographical representation of neurons containing hpGRF immunoreactivity in monkey hypothalamus. Coronal Section. Small black circles represent cell bodies. The two black areas in median eminence indicate the fiber bundles and the small dots represent the endings in contact with portal vessels. OT, optic tract; ME, median eminence; A, arcuate nucleus; VM, ventromedial nucleus; DM, dorso-medial nucleus; F, fornix; MT, mamillothalamic tract.

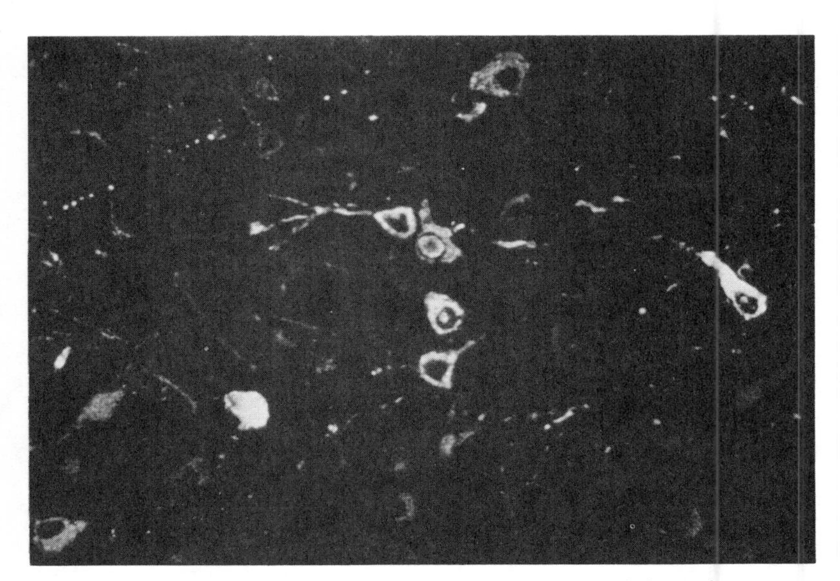

Fig. 21. GRF-immunoreactive neurons in the adult human hypothalamus (arcuate nucleus). Indirect immunoenzymatic technique. The polyclonal antiserum was raised in rabbits against synthetic hpGRF-44-NH_2. (From Bloch *et al.*, 1983*a*).

neurons producing GRF were different from those producing LRF, somatostatin, CRF and pro-opiocortin, each of them having their own specific and expected location in the hypothalamus.

Study of the ontogenesis of GRF neurons in human fetal hypothalamus (Bloch *et al.*, 1984*a*) showed that they first became detectable at the 29th week of intrauterine life, as neuroblastic cells without processes. Endings in the median eminence and fibers in the hypothalamus became detectable from the 31st week. Neurons producing GRF remain immature in aspect until birth, in contrast to other neuronal categories producing hypophysiotropic hormones, which differentiate months earlier. Since growth hormone-producing cells appear in the 8th week of fetal life in the pituitary and growth hormone becomes detectable in the blood in the 10th week, these results indicate that first stages of differentiation and activity in somatotropic cells are independent of fetal hypothalamic stimulation in the human.

Neither of the antisera against hGRF presently used did stain any structure in rat hypothalamus, showing that antigenic determinants recognized were species-specific, a finding later confirmed on the basis of the primary structure of rat hypothalamic GRF. Using antibodies against synthetic rat GRF, we have found staining in fibers and endings of rat median eminence with cell bodies located mainly in the arcuate nucleus and in ventral and dorsolateral hypothalamus. Ventromedial nucleus contains only very occasional neurons. The same investigations in rats treated neonatally with monosodium glutamate that destroyed the arcuate nucleus showed a selective disappearance of GRF neurons in the arcuate nucleus, together with disappearance of GRF fibers in the median eminence, while CRF, LRF and somatostatin fibers produced normal staining (Bloch *et al.*, 1984*b*). These results suggest that the arcuate nucleus is the source of GRF fibers projecting to the median eminence and establish monosodium glutamate-treated rats as a model for studying growth hormone secretion in rats permanently deprived of GRF stimulation. Very low abundance of GRF neurons in the human, monkey and rat ventromedial nucleus suggests that this area plays only a minor role in GRF production, contrary to what was hypothesized on the basis of electrical stimulation and destruction experiments. Since most hypothalamic GRF neurons are located in areas contiguous to the ventromedial nucleus, effects observed upon stimulation or destruction of this area must be mainly attributed to diffusion of the effect, included in the arcuate nucleus, the region that appears to be a primary source of GRF in the hypothalamus.

The search for the presence of hGRF immunoreactivity in normal primate pancreas and gut gave negative results, suggesting in particular that the pancreas is not physiologically involved in GRF production. Finally, hGRF antibodies can be used for retrospective diagnosis of putative ectopic GRF-producing tumors. Not only did we find hGRF immunoreactivity in cells of the tumor from which hGRF was isolated, but we also found it in two other tumors, one in the lung, the other in the digestive tract, both of which had produced acromegaly in the patients.

Clinical Studies with GRF

Since all current methods to stimulate secretion of GH in human subjects (i.e. hypoglycemia, L-Dopa, infusion of amino acids, etc.) are indirect, i.e. mediated by some brain center, clinical interest in a 'true' GRF, i.e. a substance that would directly stimulate secretion of GH at the pituitary level, has always been considerable. As soon as synthetic hGRF became available and clearance had been obtained from the FDA, many clinical studies were promptly started and are currently expanding. The results have not been disappointing. Intravenous injection into normal young adults, male or female, of synthetic hGRF (Rosenthal *et al.*, 1983; Chatelain *et al.*, 1983; Wood *et al.*, 1983; Gelato *et al.*, 1983) or of the fragment hGRF(1-40) (Thorner *et al.*, 1983) in doses ranging from 0.1–10 μg/kg body weight regularly produces elevation of plasma immunoreactive GH, with a peak response at 15–30 min. and lasting for 1–2 hours (Fig. 22). The response to GRF is highly specific for the secretion of GH, with no effect on plasma levels of all the other pituitary hormones or gut-peptides (Thorner *et al.*, 1983). Obviously synthetic GRF should and will replace the indirect methods currently used to assess GH-secretion or will be used in conjunction with them to finalize a diagnosis. Since all the current GH-secretion tests work indirectly, i.e. through the central nervous system, none of them being truly hypophysiotropic, they do not permit a diagnosis of GH-pituitary insufficiency as truly pituitary or hypothalamic in origin. Indeed, injection of GRF into 'hypopituitary dwarfs' has already led to the recognition of GH-secretory deficiencies truly pituitary in origin (no response to GRF) and of a second category which is best qualified as suprapituitary (most likely hypothalamic) in origin (positive response to GRF). Chronic administration of GRF should be the treatment of choice for these cases of 'hypothalamic dwarfism'. Early studies indicate that the acute

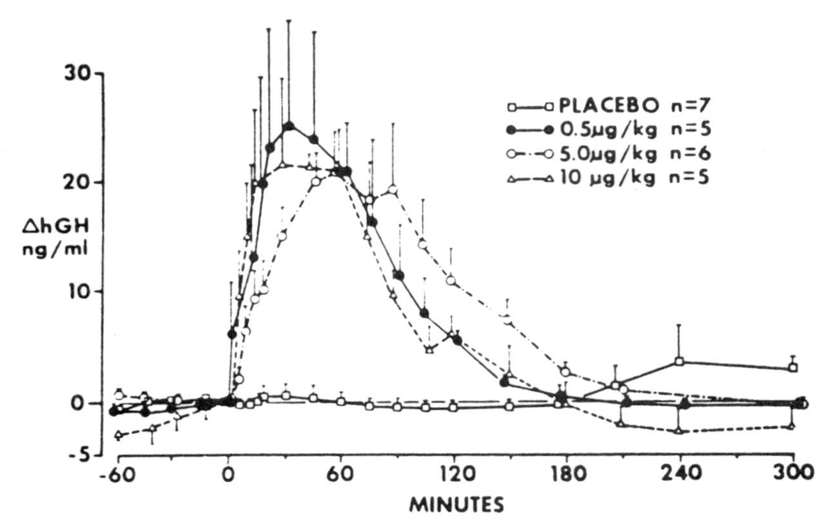

Fig. 22. Elevation of plasma GH levels in response to hpGRF in normal adult human volunteers. The response to 0.5 μg/kg, 5 μg/kg and 10 μg/kg doses of hpGRF-44 were all significantly higher than the placebo. GH levels rose within 5 min. and reached peak concentrations at 30–45 min. (0.5 μg/kg); 45–90 min. (5μg/kg) and 30–120 min. (10 μg/kg). Results are expressed in ng hGH/ml (mean ± SEM). (From Rosenthal *et al.*, 1983).

pituitary response to GRF in human subjects may be highly age-dependent, the response being quasi abolished after age 40 (Shibasaki *et al.*, 1984). This remarkable observation remains to be confirmed and explained.

Clinical interest in GRF extends over its use as a diagnostic tool and a treatment of hypothalamic dwarfism: the use of GRF can be contemplated to promote anabolism in chronic debilitating diseases, so long as the dietary intake is adequate, and to promote the healing of wounds and bone fractures. The availability of GRF with its highly specific effect in stimulating GH-secretion should permit once and for all investigation of the proposed role of GH in diabetic retinopathy. Structural analogs of GRF acting as competitive antagonists, as we have already reported in an early series (Ling and Brazeau, 1983), may be of major clinical significance in the treatment of these accidents of juvenile diabetes, something that somatostatin eventually could not offer in view of its too numerous sites of action as they came to be recognized.

Conclusions

In the space of twelve months following the characterization of human GRF from the islet cell tumor of these two acromegalic patients, one in France, the other in the USA, most of the pioneering studies on the mechanism of GRF *in vivo* and *in vitro* were described and, for some of them, completed; immunocytochemical mapping of GRF neurons in the human brain was carried out; structure-function studies were initiated; clinical trials were started and confirmed the potent GH-releasing activity of GRF in man. The effect of the peptide on specific GH-mRNA levels was described and molecular cloning was used to establish the structures of human pre-pro-GRF. The primary structure of hypothalamic GRF from several species (human, porcine, bovine, ovine, caprine and murine) was determined.

Thus all the hypothalamic releasing factors which had been postulated in the early 1950s as humoral regulators of the secretion of each pituitary hormone have now been characterized.

References

ARIMURA, A., MATSUMOTO, K., CULLER, M., TURKELSON, C., LUCIANO, M., OBARA, N., THOMAS, R., GROOT, K., SHIBATA, T., and SHIVELY, J. (1983). *Endocrinology* **112,** A291.

BARINAGA, M., YAMONOTO, G., RIVIER, C., VALE, W., EVANS, R., and ROSENFELD, M. G. (1983). *Nature* **306,** 84.

BECK, C., LARKINS, R. G., MARTIN, T. J., and BURGER, H. C. (1973). *J. Endocrinol.* **59,** 325.

BERELOWITZ, M., SZAKO, M., FROHMAN, L. A., FIRESTONE, S. and CHU, L. (1981). *Science* **212,** 1279.

BLEULER, M. (1951). *J. Nerv. Ment. Dis.* **113,** 497.

BLOCH, B., BRAZEAU, P., BLOOM, F., and LING, N. (1983*a*). *Neurosci. Lett.* **37,** 23–8.

BLOCH, B., BRAZEAU, P., LING, N., BÖHLEN, P., ESCH, F., WEHRENBERG, W. B., BENOIT, R., BLOOM, F. and GUILLEMIN, R. (1983*b*). *Nature* **301,** 607.

BLOCH, B., GAILLARD, R. C., BRAZEAU, P., LIN, H. D., and LING, N. (1984*a*). *Reg. Peptides* **8,** 21–31.

BLOCH, B., LING, N., BENOIT, R., WEHRENBERG, W. B., and GUILLEMIN, R. (1984*b*). *Nature* **301,** 272–73.

BÖHLEN, P., and MELLET, M. (1979). *Anal. Biochem.* **94,** 313–21.

BÖHLEN, P., and SCHROEDER, R. (1982). *Anal. Biochem.* **126,** 144.

BÖHLEN, P., THORNER, M., CRONIN, M., SHIVELY, J., and SCHEITHAUER, B. (1982). *Endocrinology* **110,** A540.

BÖHLEN, P., BRAZEAU, P., ESCH, F., LING, N., WEHRENBERG, W. B., and GUILLEMIN, R. (1983*a*). *Reg. Peptides* **6,** 343.

BÖHLEN, P., ESCH, F., BRAZEAU, P., LING, N., and GUILLEMIN, R. (1983*b*). *Biochem. Biophys. Res. Commun.* **116,** 726–34.

BÖHLEN, P., WEHRENBERG, W. B., ESCH, F., LING, N., BRAZEAU, P., and GUILLEMIN, R. (1984). *Biochem. Biophys. Res. Commun.* **125,** 1005–12.

BOYD, A., SANCHEZ-FRANCO, E., SPENCER, E., PATEL, Y. C., JACKSON, I. M. D., and REICHLIN, S. (1978). *Endocrinology* **103,** 1075.

BRAZEAU, P., VALE, W., BURGUS, R., LING, N., BUTCHER, M., RIVIER, J., and GUILLEMIN, R. (1973). *Science* **179,** 77.

BRAZEAU, P., LING, N., ESCH, F., BÖHLEN, P., BENOIT, R., and GUILLEMIN, R. (1981*a*). *Reg. Peptides* **1,** 255.

BRAZEAU, P., BÖHLEN, P., LING, N., ESCH, F., BENOIT, R., and GUILLEMIN, R. (1981*b*). *Endocrinology* **108,** A837.

BRAZEAU, P., LING, N., BÖHLEN, P., ESCH, F., YING, S.-Y., and GUILLEMIN, R. (1982*a*). *Proc. Natl. Acad. Sci. USA* **79,** 7909.

BRAZEAU, P., GUILLEMIN, R., LING, N., VAN WYK, J., and HUMBEL, R. (1982*b*). *C.R. Acad. Sci.* (Paris) **295,** 651.

Brazeau, P., Böhlen, P., Esch, F., Ling, N., Wehrenberg, W. B., and Guillemin, R. (1984). *Biochem. Biophys. Res. Commun.* **125,** 606–14.

Burgus, R., Dunn, T., Desiderio, D., and Guillemin, R. (1969). *C.R. Acad. Sci.* (Paris) **269,** 1870–73.

Burgus, R., Butcher, M., Ling, N., Monahan, M., Rivier, J., Fellows, R., Amoss, M., Blackwell, R., Vale, W., and Guillemin, R. (1971). *C.R. Acad. Sci.* (Paris) **273,** 1611–13.

Caplan, R. H., Koob, L., Abellera, R. M., Pagliara, A. S., Kovacs, K., and Randall, R. V. (1978). *Amer. J. Med.* **64,** 874.

Chatelain, P., Cohen, H., Sassolas, G., Exclerc, J. L., Ruitton, A., Cohen, R., Claustrat, B., Laporte, S., Laferre, B., Elcharfi, A., Ferry, S., and Guillemin, R. (1983). *Ann. Endocr.* (Paris) **44,** A25.

Cronin, M. J., Rogol, A. D., Dabney, L. G., and Thorner, M. O. (1982). *J. Clin. Endocrinol. Metab.* **55,** 381.

De Lean, A., Munson, P. J., and Rodbard, D. (1978). *Am. J. Physiol.* **235** (2), E97.

Dhariwal, A. P. S., Krulich, L., Katz, S. H., and McCann, S. M. (1965). *Endocrinology* **77,** 932.

Ducommun, P., Vale, W., Sakiz, E., and Guillemin, R. (1967). *Endocrinology* **80,** 953.

du Vigneaud, V. (1952). 2nd Cong. intern. biochim., Chim. biol. IV, *Symposium hormones protéiques et dérivées des protéines* (Paris).

Ehlers, C. L., Henriksen, S., Reed, T. K., and Bloom, F. E. (1985). *Neuroendocrinology* (to be published).

Esch, F. S., Böhlen, P., Ling, N. C., Brazeau, P. E., Wehrenberg, W. B., Thorner, M. O., Cronin, M. J., and Guillemin, R. (1982). *Biochem. Biophys. Res. Commun.* **109,** 152.

Esch, F., Böhlen, P., Ling, N., Brazeau, P., and Guillemin, R. (1983). *Biochem. Biophys. Res. Commun.* **117,** 772–79.

Faden, V. B., Huston, J., Munson, P., and Rodbard, D. (1980). *Logit-log analysis of radioimmunoassay.* NICHD RRB NIH.

Frohman, L. A., Szabo, M., Berelowitz, M., and Stachura, M. E. (1980). *J. Clin. Invest.* **65,** 43.

Gelato, M. C., Pescovitz, O., Cassola, F., Loriaux, L., and Merrian, G. (1983). *J. Clin. Endocr. Metab.* **57,** 674.

Gick, G. G., Zeytinoglu, F. N., Esch, F. S., and Bancroft, F. C. (1983). *Endocrinology* **112,** A295.

Green, J. D. (1951). *Amer. J. Anat.* **88,** 225–312.

Gubler, U., Monahan, J. J., Lomedico, P. T., Bhatt, R. S., Collier, K. J., Hoffman, B. J., Böhlen, P., Esch, F., Ling, N., Zeytin, F., Bra-

ZEAU, P., POONIAN, M. S., and GAGE, L. P. (1983). *Proc. Natl. Acad. Sci. USA* **80,** 4311.

GUILLEMIN, R., LING, N., and VARGO, T. (1977). *Biochem. Biophys. Res. Commun.* **77,** 361.

GUILLEMIN, R., BRAZEAU, P., BÖHLEN, P., ESCH, F., LING, N., and WEHRENBERG, W. (1982). *Science* **218,** 585.

GUILLEMIN, R., BRAZEAU, P., BÖHLEN, P., ESCH, F., LING, N., WEHRENBERG, W. B., BLOCH, B., MOUGIN, C., ZEYTIN, F., and BAIRD, A. (1984). In *Recent Progress in Hormone Research*, Vol. 40. R. Greep (ed.), Academic Press, NY, 233–99.

HARRIS, G. W. (1955). *Neural Control of the Pituitary Gland,* Vol. 1. Edward Arnold Ltd. Publ. London.

HESS, W. R. (1948). *Die funktionelle Organization des vegetativen Nervensystems*. Benno Schwabe, Publ., Basel.

HEWICK, R. M., HUNKAPILLER, M. W., HOOD, L. E., and DREYER, W. J. (1981). *J. Bio. Chem.* **15,** 7990.

HUNKAPILLER, M. W., and HOOD, L. E. (1983). *Methods in Enzymol.* (Hirs and Timasheff eds.), Academic Press, New York, **91,** 487.

HUNKAPILLER, M. W., HEWICK, R. W., DREYER, W. J., and HOOD, L. E. (1983). *Methods in Enzymol.* (Hirs and Timasheff eds.), Academic Press, New York, **91,** 399.

JOHANSSON, K., CURRIE, B., FOLKERS, K., and BOWERS, C. Y. (1974). *Biochem. Biophys. Res. Commun.* **60,** 610.

KADIKANI, H., FURUTANI, Y., TARAHASHI, H., NODA, M., MORIMOTO, Y., HIROSE, T., ASAI, M., INAYAMA, S., NAKANISHI, S., and NUMA, S. (1982). *Nature* **298,** 245.

KRULICH, L., DHARIWAL, A. P. S., and MCCANN, S. M. (1968). *Endocrinology* **83,** 783.

LEVESTON, A., MCKEEL, D. W., BUCKLEY, P. J., DESCHRYVER, K., GREIDER, M. H., JAFFE, B. M., and DAUGHADAY, W. H. (1981). *J. Clin. Endocrinol. Metab.* **53,** 682.

LEWIN, M. J. M., REYL-DESMARS, F., and LING, N. (1983). *Proc. Natl. Acad. Sci.* **80,** 6538–41.

LING, N., and BRAZEAU, P. (1983). *Endocrinology* **112,** A295.

LING, N., ESCH, F., DAVIS, D., MERCADO, M., REGNO, M., BÖHLEN, P., BRAZEAU, P., and GUILLEMIN, R. (1980). *Biochem. Biophys. Res. Commun.* **95,** 945.

LING, N., ESCH, F., BÖHLEN, P., BRAZEAU, P., WEHRENBERG, W. B., and GUILLEMIN, R. (1984). *Proc. Natl. Acad. Sci. USA* **81,** 4202.

LUBEN, R. A., BRAZEAU, P., BÖHLEN, P., and GUILLEMIN, R. (1982). *Science* **218,** 887.

Malacara, J. M., Valverde, R. C., Reichlin, S., and Bollinger, J. (1972). *Endocrinology* **91,** 1189.

Mandell, A. J. (1983). In *Synergetics of Brain*, Basar, E., Flohr, H., Haken, H., Mandell, A. J. (eds.), Springer-Verlag, Berlin, 365–76.

Matsuo, H., Baba, Y., Nair, R. M. G., Arimura, A., and Schally, A. V. (1971). *Biochem. Biophys. Res. Commun.* **43,** 1334–39.

Mayo, K. E., Vale, W., Rivier, J., Rosenfeld, M. G., and Evans, R. M. (1983). *Nature* **306,** 86.

Minamino, N., Kangawa, K., Fukuda, A., and Matsuo, H. (1980). *Biochem. Biophys. Res. Commun.* **95,** 1475.

Morel, G., Mesguich, P., Dubois, M. P., and Dubois, P. M. (1984). *Neuroendocrinology* **38,** 123–33.

Nair, R. M. G., Barrett, F. J., Bowers, C. Y., and Schally, A. V. (1970). *Biochemistry* **9,** 1103.

Nair, R. M. G., de Villier, C., Barnes, M., Antalis, J., and Wilbur, D. L. (1978). *Endocrinology* **103,** 112.

Reyl, F. J., and Lewin, M. J. M. (1981). *Proc. Natl. Acad. Sci. USA* **79,** 978.

Rivier, J. (1978). *J. Liq. Chromatogr.* **1,** 343–66.

Rivier, J., Spiess, J., Thorner, M. O., and Vale, W. (1982). *Nature* **300,** 276.

Rodbard, D. (1974). *Clin. Chem.* **20,** 1255.

Rosenthal, S., Schriock, E., Kaplan, S., Guillemin, R., and Grumbach, M. (1983). *J. Clin. Endocrinol. Metab.* **57,** 677.

Rubinstein, M., Stein, S., Gerber, L., and Udenfriend, S. (1977). *Proc. Natl. Acad. Sci. USA* **74,** 3052.

Sachar, E. J. (1975). In *Topics in Neuroendocrinology*, (E. J. Sachar, ed.), p. 135. Grune and Stratton, New York.

Sakiz, E. (1964). *Excerpta Medica Internat. Congress Series no.* **83,** 225.

Sassolas, G., Rousset, H., Cohen, R., and Chatelain, P. (1983). *C. R. Acad. Sci.* (Paris) **296,** 527.

Schally, A. V., Sawano, S., Arimura, A., Barrett, J. F., Wakabayashi, I., and Bowers, C. Y. (1969). *Endocrinology* **84,** 1493.

Schally, A. V., Baba, Y., Nair, R. M. G., and Bennett, C. D. (1971). *J. Biol. Chem.* **246,** 6647.

Selye, H. (1946). *J. Clin. Endocrinol.* **6,** 117.

Sherrington, C. S. (1906). *The Integrative Action of the Hormone System.* Yale Univ. Press, New Haven.

Shibasaki, T., Shizume, K., Nakahara, M., Masuda, A., Jibiki, K., Demura, H., Wakabayashi, I., and Ling, N. (1984*a*). *J. Clin. Endocrinol.* **58,** 212.

Shibasaki, T., Shizume, K., Masuda, A., Nakahara, M., Hizuke, N.,

MIYAKAWA, M., TAKANO, K., DEMURA, W., WAKABAYASHI, I. and LING, N. (1984*b*). *J. Clin. Endocrinol.* **58,** 215.

SINHA, Y. N., SELBY, F. W., LEWIS, U. J., and VANDERLAAN, W. P. (1972). *Endocrinology* **91,** 784.

SPIESS, J., RIVIER, J., and VALE, W. (1983). *Nature* **303,** 532.

STACHURA, M. E., DHARIWAL, A. P. S., and FROHMAN, L. A. (1972). *Endocrinology* **91,** 1071.

SYKES, J. E., and LOWRY, P. J. (1983). *Biochem. J.* **209,** 643.

TANNENBAUM, G. S., and MARTIN, J. B. (1976). *Endocrinology* **98,** 562–70.

TANNENBAUM, G. S., EPELBAUM, J., COLLE, E., BRAZEAU, P., and MARTIN, J. B. (1978). *Endocrinology* **102,** 1909.

TASHJIAN, A. H., BAROWSKY, N. J., and JENSEN, D. K. (1971). *Biochem. Biophys. Res. Commun.* **43,** 516–23.

THORNER, M. O., PERRYMAN, R. L., CRONIN, M. J., DRAZNIN, M., JOHANSON, A., ROGOL, A. D., JANE, J., RUDOLF, L., HORVATH, E., KOVACS, K., and VALE, W. (1982). *Clin. Res.* **30,** 555A.

THORNER, M., SPIESS, J., VANCE, M., ROGOL, A., KAISER, D., WEBSTER, J., RIVIER, J., BORGES, J., BLOOM, S., CRONIN, M., EVANS, W., MACLEOD, R., and VALE, W. (1983). *Lancet* **1,** 24.

UZZAFAR, M. S., MELLINGER, R. C., FINE, G., SZABO, M., and FROHMAN, L. A. (1979). *J. Clin. Endocrinol. Metab.* **48,** 66.

VALE, W., BURGUS, R., and GUILLEMIN, R. (1968). *Neuroendocrinology* **3,** 34.

VALE, W., SPIESS, J., RIVIER, C., and RIVIER, J. (1981). *Science* **213,** 1394–97.

VEBER, D. F., BENNETT, C. D., MILKOWSKI, J. D., GAL, G., DENKEWALTER, R. G., and HIRSCHMANN, R. (1971). *Biochem. Biophys. Res. Commun.* **45,** 235.

WEHRENBERG, W. B., LING, N., BRAZEAU, P., ESCH, F., BÖHLEN, P., BAIRD, A., YING, S., and GUILLEMIN, R. (1982*a*). *Biochem. Biophys. Res. Commun.* **109,** 382.

WEHRENBERG, W. B., LING, N., BÖHLEN, P., ESCH, F., BRAZEAU, P., and GUILLEMIN, R. (1982*b*). *Biochem. Biophys. Res. Commun.* **109,** 562.

WEHRENBERG, W. B., BRAZEAU, P., LUBEN, R., BÖHLEN, P., and GUILLEMIN, R. (1982*c*). *Endocrinology* **111,** 2147.

WEHRENBERG, W. B., BRAZEAU, P., LUBEN, R., LING, N., and GUILLEMIN, R. (1983*a*). *Neuroendocrinology* **36,** 489.

WEHRENBERG, W. B., BAIRD, A., YING, S.-Y., RIVIER, C., LING, N., and GUILLEMIN, R. (1984*a*). *Endocrinology* **114,** 1995–2001.

WEHRENBERG, W. B., BLOCH, B., ZHANG, C.-L., LING, N., and GUILLEMIN, R. (1984*b*). *Reg. Peptides* **8,** 1–8.

WHITE, B. A., and BANCROFT, F. C. (1983). *J. Biol. Chem.* **258,** 4618.

WILBER, J. F., NAGEL, T., and WHITE, W. F. (1971). *Endocrinology* **89,** 1419.

WILSON, M. C., STEINER, A. L., DHARIWAL, A. P. S., and PEAKE, G. T. (1974). *Neuroendocrinology* **15,** 313--27.

WOOD, S. M., CHING, J. L. C., ADAMS, E. F., WEBSTER, J. D., JOPLING, F., MASHITER, K., and BLOOM, S. R. (1983). *Brit. Med. J.* **286,** 1687.

YUDAEV, N. A., UTESHEVA, Z. F., NOVIKOVA, T. E., SHVACHKI, Y. P., and SMIRNOVA, A. P. (1973). *Dan. SSSR.* **210,** 731.

ZEYTINOGLU, F., and BRAZEAU, P. (1984). *Biochem. Biophys. Res. Commun.* **123,** 497–506.

Index